Algebra I
Station Activities
for Common Core Standards

1   2   3   4   5   6   7   8   9   10
ISBN 978-0-8251-6786-7
Copyright © 2011
J. Weston Walch, Publisher
Portland, ME 04103
www.walch.com
Printed in the United States of America

WALCH EDUCATION®

Table of Contents

Algebra I Station Activities for Common Core State Standards

Standards Correlations

The standards correlations below support the implementation of the Common Core Standards. *Algebra I Station Activities for Common Core State Standards* includes station activity sets for the Common Core areas of Number and Quantity, Algebra, Functions, and Statistics and Probability. This table provides a listing of the available station activities organized by Common Core standard.

The left column lists the standard codes. The first letter of the code represents the Common Core area. The area letter is followed by a dash and the initials of the domain name, which is then followed by the standard number. The middle column lists the title of the station activity set that corresponds to the standard, and the right column lists the page number where the station activity set can be found.

The table indicates the standards that are heavily addressed in the station sets. If there are other standards that are addressed within the set, they can be found on the first page of each set.

Standard	Set title	Page number
N-Q.1.	Ratios and Proportions	15
N-VM.7.	Matrices	1
N-VM.8.	Matrices	1
A-SSE.2.	Simplifying Radical Expressions with Variables	28
A-SSE.2.	Operations with Radicals and Variables	39
A-SSE.2.	Factoring Polynomials	53
A-SSE.3.	Factoring Polynomials	53
A-CED.1.	Ratios and Proportions	15
A-CED.1.	Solving Linear Equations	107
A-CED.2.	Graphing Linear Equations/Solving Using Graphs	81
A-CED.2.	Writing Linear Equations	99
A-CED.2.	Real-World Situation Graphs	119
A-CED.3.	Solving Systems of Inequalities	204
A-CED.4.	Literal Equations	70
A-REI.3.	Ratios and Proportions	15
A-REI.3.	Solving Linear Equations	107
A-REI.3.	One-Variable Inequalities	132
A-REI.4.	Solving Quadratic Equations by Finding Square Roots	221
A-REI.4.	Solving Quadratic Equations Using the Quadratic Formula	234
A-REI.5.	Solving 2-by-2 Systems by Elimination	180
A-REI.5.	Using Systems in Applications	191
A-REI.6.	Solving 2-by-2 Systems by Graphing	159
A-REI.6.	Solving 2-by-2 Systems by Substitution	170

(continued)

Standards Correlations

Algebra I Station Activities for Common Core State Standards
© 2011 Walch Education

Introduction

Algebra I Station Activities for Common Core State Standards includes a collection of station-based activities to provide students with opportunities to practice and apply the mathematical skills and concepts they are learning. It contains several sets of activities for each of the following Common Core High School Mathematics areas: Number and Quantity; Algebra (Seeing Structure in Expressions; Creating Equations; Reasoning with Equations and Inequalities; Arithmetic with Polynomials and Rational Expressions); Functions (Interpreting Functions); and Statistics and Probability. You may use these activities as a complement to your regular lessons or in place of your regular lessons, if formative assessment suggests students have the basic concepts but need practice. The debriefing discussions after each set of activities provide an important opportunity to help students reflect on their experiences and synthesize their thinking. It also provides an additional opportunity for ongoing, informal assessment to inform instructional planning.

Implementation Guide

The following guidelines will help you prepare for and use the activity sets in this book.

Setting Up the Stations

Each activity set consists of four or more stations. Set up each station at a desk, or at several desks pushed together, with enough chairs for a small group of students. Place a card with the number of the station on the desk. Each station should also contain the materials specified in the teacher's notes, and a stack of student activity sheets (one copy per student). Place the required materials (as listed) at each station.

When a group of students arrives at a station, each student should take one of the activity sheets to record the group's work. Although students should work together to develop one set of answers for the entire group, each student should record the answers on his or her own activity sheet. This helps keep students engaged in the activity and gives each student a record of the activity for future reference.

Forming Groups of Students

All activity sets consist of four stations. You might divide the class into four groups by having students count off from 1 to 4. If you have a large class and want to have students working in small groups, you might set up two identical sets of stations, labeled A and B. In this way, the class can be divided into eight groups, with each group of students rotating through the "A" stations or "B" stations.

Introduction

Assigning Roles to Students

Students often work most productively in groups when each student has an assigned role. You may want to assign roles to students when they are assigned to groups and change the roles occasionally. Some possible roles are as follows:

- Reader—reads the steps of the activity aloud
- Facilitator—makes sure that each student in the group has a chance to speak and pose questions; also makes sure that each student agrees on each answer before it is written down
- Materials Manager—handles the materials at the station and makes sure the materials are put back in place at the end of the activity
- Timekeeper—tracks the group's progress to ensure that the activity is completed in the allotted time
- Spokesperson—speaks for the group during the debriefing session after the activities

Timing the Activities

The activities in this book are designed to take approximately 15 minutes per station. Therefore, you might plan on having groups change stations every 15 minutes, with a two-minute interval for moving from one station to the next. It is helpful to give students a "5-minute warning" before it is time to change stations.

Since the activity sets consist of four stations, the above timeframe means that it will take about an hour and 10 minutes for groups to work through all stations. If this is followed by a 20-minute class discussion as described on the next page, an entire activity set can be completed in about 90 minutes.

Guidelines for Students

Before starting the first activity set, you may want to review the following "ground rules" with students. You might also post the rules in the classroom.

- All students in a group should agree on each answer before it is written down. If there is a disagreement within the group, discuss it with one another.
- You can ask your teacher a question only if everyone in the group has the same question.
- If you finish early, work together to write problems of your own that are similar to the ones on the student activity sheet.
- Leave the station exactly as you found it. All materials should be in the same place and in the same condition as when you arrived.

Introduction

Debriefing the Activities

After each group has rotated through every station, bring students together for a brief class discussion. At this time you might have the groups' spokespersons pose any questions they had about the activities. Before responding, ask if students in other groups encountered the same difficulty or if they have a response to the question. The class discussion is also a good time to reinforce the essential ideas of the activities. The questions that are provided in the teacher's notes for each activity set can serve as a guide to initiating this type of discussion.

You may want to collect the student activity sheets before beginning the class discussion. However, it can be beneficial to collect the sheets afterward so that students can refer to them during the discussion. This also gives students a chance to revisit and refine their work based on the debriefing session.

Guide to Common Core Standards Annotation

As you use this book, you will come across annotation symbols included with the Common Core standards for several station activities. The following descriptions of these annotation symbols are verbatim from the Common Core State Standards Initiative Web site, at www.corestandards.org.

Symbol: ★

Denotes: Modeling Standards

Modeling is best interpreted not as a collection of isolated topics but rather in relation to other standards. Making mathematical models is a Standard for Mathematical Practice, and specific modeling standards appear throughout the high school standards indicated by a star symbol (★).

From http://www.corestandards.org/the-standards/mathematics/high-school-modeling/introduction/

Symbol: (+)

Denotes: College and Career Readiness Standards

The evidence concerning college and career readiness shows clearly that the knowledge, skills, and practices important for readiness include a great deal of mathematics prior to the boundary defined by (+) symbols in these standards.

From http://www.corestandards.org/the-standards/mathematics/note-on-courses-and-transitions/courses-and-transitions/

Introduction

Materials List

Class Sets

- calculators
- rulers

Station Sets

- at least 40 red, 25 blue, 25 green, and 20 yellow algebra tiles
- measuring stick
- graphing calculators
- at least 24 green, 16 yellow, 1 red, and 1 blue marbles
- bag to hold marbles
- spaghetti noodles
- slips of paper with $<$, $>$, $\leq$, and $\geq$ written on them
- deck of playing cards that contains only the numbers 2–10
- 3 pieces of red yarn, 3 pieces of blue yarn
- a fair coin

Ongoing Use

- highlighters (yellow specifically)
- index cards (prepared according to specifications in teacher notes for many of the station activities)
- number cubes
- graph paper
- pencils
- tape

Number and Quantity

Set 1: Matrices

Goal: To provide opportunities for students to develop concepts and skills related to adding and subtracting matrices, scalar multiplication, and matrix multiplication

Common Core Standards

Number and Quantity: Vector and Matrix Quantities

Perform operations on matrices and use matrices in applications.

N-VM.7. (+) Multiply matrices by scalars to produce new matrices; e.g., as when all of the payoffs in a game are doubled.

N-VM.8. (+) Add, subtract, and multiply matrices of appropriate dimensions.

Student Activities Overview and Answer Key

Station 1

Students are given 18 index cards with the following integers written on them: –10, –7, –3, –1, 0, 1, 2, 4, 8, 9, 10, 12, 15, 20, 22, 25, 40, 100. Students draw the index cards from a pile to fill in the matrices. Then they work together to perform matrix addition. Students explain why matrix addition is not possible on matrices that are not the same size.

Answers

1. Answers will vary depending on which cards students draw. Possible answers:

$$\begin{bmatrix} -10 & -7 \\ 10 & 12 \end{bmatrix} + \begin{bmatrix} 15 & 22 \\ 8 & 9 \end{bmatrix} = \begin{bmatrix} 5 & 15 \\ 18 & 21 \end{bmatrix}$$

2. Answers will vary. Possible answers:

$$\begin{bmatrix} -3 & 8 & -10 \\ 0 & 4 & 40 \\ 1 & 9 & 100 \end{bmatrix} + \begin{bmatrix} -1 & 20 & -7 \\ 25 & 15 & 10 \\ 12 & 2 & 22 \end{bmatrix} = \begin{bmatrix} -4 & 28 & -17 \\ 25 & 19 & 50 \\ 13 & 11 & 122 \end{bmatrix}$$

3. $$\begin{bmatrix} 1 & 40 \\ 0 & -12 \end{bmatrix} + \begin{bmatrix} 100 & -5 \\ 88 & -7 \end{bmatrix} = \begin{bmatrix} 101 & 35 \\ 88 & -19 \end{bmatrix}$$

4. $$\begin{bmatrix} 14 & 0 & -14 \\ -6 & -7 & 12 \\ 18 & 24 & 88 \end{bmatrix} + \begin{bmatrix} 1 & 0 & -10 \\ 14 & 49 & 56 \\ -22 & -28 & 0 \end{bmatrix} = \begin{bmatrix} 15 & 0 & -24 \\ 8 & 42 & 68 \\ -4 & -4 & 88 \end{bmatrix}$$

5. Matrix addition is not possible because the matrices are different sizes.

Station 2

Students are given 18 index cards with the following integers written on them: –12, –11, –4, –1, 0, 1, 2, 4, 7, 9, 14, 16, 21, 30, 31, 38, 75, 90. Students draw the index cards from a pile to fill in the matrices. Then they work together to perform matrix subtraction. Students explain why matrix subtraction is not possible on matrices that are not the same size.

Answers

1. Answers will vary depending on the cards drawn. Possible answers:

$$\begin{bmatrix} 9 & 31 \\ -4 & 90 \end{bmatrix} - \begin{bmatrix} -12 & 16 \\ 4 & 7 \end{bmatrix} = \begin{bmatrix} 21 & 15 \\ -8 & 83 \end{bmatrix}$$

2. Answers will vary. Possible answers:

$$\begin{bmatrix} -12 & 4 & 1 \\ 14 & 0 & 21 \\ 9 & 31 & -11 \end{bmatrix} - \begin{bmatrix} 7 & 2 & -1 \\ 16 & -4 & 30 \\ 38 & 90 & 75 \end{bmatrix} = \begin{bmatrix} -19 & 2 & 2 \\ -2 & 4 & -9 \\ -29 & -59 & -86 \end{bmatrix}$$

3.
$$\begin{bmatrix} 3 & -5 \\ 44 & 0 \end{bmatrix} - \begin{bmatrix} 1 & -25 \\ 12 & -18 \end{bmatrix} = \begin{bmatrix} 2 & 20 \\ 32 & 18 \end{bmatrix}$$

4.
$$\begin{bmatrix} 0 & 22 & -11 \\ 15 & 44 & 72 \\ -16 & 8 & -1 \end{bmatrix} - \begin{bmatrix} 0 & 8 & -4 \\ -14 & 7 & 12 \\ 22 & 39 & -21 \end{bmatrix} = \begin{bmatrix} 0 & 14 & -7 \\ 29 & 37 & 60 \\ -38 & -31 & 20 \end{bmatrix}$$

5. Even though the matrices have the same amount of numbers, they are different sizes. Therefore, it is impossible to perform matrix subtraction.

Station 3

Students are given 12 index cards with the following scalars and matrices on them:

Scalars: –2, –3, 1, 3, 4, 5, 8, 12

Matrices: $\begin{bmatrix} 10 & -4 & 5 \end{bmatrix}$, $\begin{bmatrix} 12 & 7 \\ 8 & 3 \end{bmatrix}$, $\begin{bmatrix} 10 & 2 & 3 \\ -2 & 4 & 12 \\ 7 & 11 & 3 \end{bmatrix}$, $\begin{bmatrix} -5 & -8 \\ -9 & -3 \end{bmatrix}$

Students place the scalars in one pile and the matrices in another pile. One student draws a card from each pile. The students work together to perform scalar multiplication based on the two cards drawn. Students find scalar multiples of a matrix.

Answers

1. Answers will vary depending on the cards drawn. Possible answers:

$$-3\begin{bmatrix} 10 & -4 & 5 \end{bmatrix} = \begin{bmatrix} -30 & 12 & -15 \end{bmatrix}$$

2. Answers will vary. Possible answers:

$$5\begin{bmatrix} 12 & 7 \\ 8 & 3 \end{bmatrix} = \begin{bmatrix} 60 & 35 \\ 40 & 15 \end{bmatrix}$$

3. Answers will vary. Possible answers:

$$1\begin{bmatrix} 10 & 2 & 3 \\ -2 & 4 & 12 \\ 7 & 11 & 3 \end{bmatrix} = \begin{bmatrix} 10 & 2 & 3 \\ -2 & 4 & 12 \\ 7 & 11 & 3 \end{bmatrix}$$

4. Scalar = −5; the numbers in the second matrix are −5 times the numbers in the first matrix.

5. Original matrix: $\begin{bmatrix} -1 & 9 \\ -5 & 4 \end{bmatrix}$

Station 4

Students will be given a number cube. As a group, they will determine how to perform matrix multiplication based on a given problem and answer. Then they will roll the dice to populate matrices and perform matrix multiplication.

Answers

1. If you have matrices *A* and *B*, then *AB* = multiplying the rows in matrix *A* by the columns in matrix *B* and then finding their sum. The number of columns in *A* must equal the number of rows in *B* in order for the matrix multiplication to be defined.

2. Answers will vary. Possible answers:

$$\begin{bmatrix} 4 & 1 \end{bmatrix}\begin{bmatrix} 3 \\ 6 \end{bmatrix} = \begin{bmatrix} 4(3) + 1(6) \end{bmatrix} = \begin{bmatrix} 12 + 6 \end{bmatrix} = \begin{bmatrix} 18 \end{bmatrix}$$

3. Answers will vary. Possible answers:

$$\begin{bmatrix} 2 & 3 & 1 \\ 6 & 1 & 4 \end{bmatrix}\begin{bmatrix} 3 & 5 \\ 2 & 5 \\ 4 & 6 \end{bmatrix} = \begin{bmatrix} 2(3) + (3)(2) + (1)(4) & 2(5) + (3)(5) + (1)(6) \\ 6(3) + (1)(2) + (4)(4) & 6(5) + (1)(5) + (4)(6) \end{bmatrix} = \begin{bmatrix} 16 & 31 \\ 36 & 59 \end{bmatrix}$$

Materials List/Setup

Station 1 18 index cards with the following integers written on them:

−10, −7, −3, −1, 0, 1, 2, 4, 8, 9, 10, 12, 15, 20, 22, 25, 40, 100

Station 2 18 index cards with the following integers written on them:

−12, −11, −4, −1, 0, 1, 2, 4, 7, 9, 14, 16, 21, 30, 31, 38, 75, 90

Station 3 12 index cards with the following scalars and matrices on them:

Scalars: −2, −3, 1, 3, 4, 5, 8, 12

Matrices: $\begin{bmatrix} 10 & -4 & 5 \end{bmatrix}$, $\begin{bmatrix} 12 & 7 \\ 8 & 3 \end{bmatrix}$, $\begin{bmatrix} 10 & 2 & 3 \\ -2 & 4 & 12 \\ 7 & 11 & 3 \end{bmatrix}$, $\begin{bmatrix} -5 & -8 \\ -9 & -3 \end{bmatrix}$

Station 4 number cube

Discussion Guide

To support students in reflecting on the activities and to gather formative information about student learning, use the following prompts to facilitate a class discussion to "debrief" the station activities.

Prompts/Questions

1. How do you perform matrix addition for two matrices $A + B$?

2. How do you perform matrix subtraction for two matrices $A - B$?

3. Why does the size of each matrix matter when performing addition or subtraction?

4. How do you perform scalar multiplication with matrices?

5. How do you perform matrix multiplication for two matrices, A and B?

6. What are real-world examples of matrices?

Think, Pair, Share

Have students jot down their own responses to questions, then discuss with a partner (who was not in their station group), and then discuss as a whole class.

Suggested Appropriate Responses

1. Add the element in row 1, column 1 of matrix A with the corresponding element in row 1, column 1 of matrix B. Repeat this process for the remaining corresponding elements.

2. Subtract the element in row 1, column 1 of matrix B from the corresponding element in row 1, column 1 of matrix A. Repeat this process for the remaining corresponding elements.

3. Matrix addition and subtraction relies on corresponding elements. Therefore, the matrices must be the same size to provide matches for all the corresponding elements.

4. Multiply the scalar by each element in the matrix.

5. If you have matrices A and B, then AB = multiplying the rows in matrix A by the columns in matrix B and then finding their sum. The number of columns in A must equal the number of rows in B in order for the matrix multiplication to be defined.

6. When you organize data in columns and rows, it is considered a matrix. Examples include students and their test scores, growth rates of plants, and populations of states.

Possible Misunderstandings/Mistakes

- Incorrectly subtracting negative numbers during matrix subtraction
- Subtracting the matrices in the wrong order
- Trying to add or subtract two matrices that are not the same size
- Not keeping track of which row is multiplied by which column
- Mistakenly finding the sum instead of the product during matrix multiplication
- Not keeping track of which steps they take to find their answer
- Not keeping track of the signs of the numbers

Number and Quantity
Set 1: Matrices

Station 1

You will be given 18 index cards with the following numbers written on them:

−10, −7, −3, −1, 0, 1, 2, 4, 8, 9, 10, 12, 15, 20, 22, 25, 40, 100

Place the index cards in a pile. Take turns drawing a card from the top of the pile. Write the number from each card in one of the boxes in the matrix below. Repeat this process until all boxes of each matrix contain a number.

1. Perform matrix addition on the matrices above. Show your work in the space below.

2. Place the cards back in the pile and shuffle. Take turns drawing one card at time. Fill in the boxes of the matrices with the numbers from the cards, and perform matrix addition.

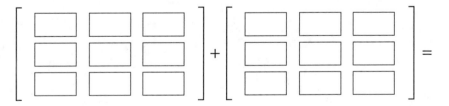

continued

Number and Quantity
Set 1: Matrices

As a group, perform matrix addition on the following matrices. If matrix addition is not possible, explain why. Discuss your answers.

3. $\begin{bmatrix} 1 & 40 \\ 0 & -12 \end{bmatrix} + \begin{bmatrix} 100 & -5 \\ 88 & -7 \end{bmatrix} =$

4. $\begin{bmatrix} 14 & 0 & -14 \\ -6 & -7 & 12 \\ 18 & 24 & 88 \end{bmatrix} + \begin{bmatrix} 1 & 0 & -10 \\ 14 & 49 & 56 \\ -22 & -28 & 0 \end{bmatrix} =$

5. $\begin{bmatrix} 2 \\ 25 \end{bmatrix} + \begin{bmatrix} 14 & 25 \\ 35 & 15 \end{bmatrix} =$

Algebra I Station Activities for Common Core State Standards

Number and Quantity
Set 1: Matrices

Station 2

You will be given 18 index cards with the following numbers written on them:

−12, −11, −4, −1, 0, 1, 2, 4, 7, 9, 14, 16, 21, 30, 31, 38, 75, 90

Place the index cards in a pile. Take turns drawing a card from the top of the pile. Write the number from each card in one of the boxes in the matrix below. Repeat this process until all boxes of each matrix contain a number.

1. Perform matrix subtraction on the matrices above. Show your work in the space below.

2. Place the cards back in the pile and shuffle. Take turns drawing one card at time. Fill in the boxes of the matrices with the numbers from the cards, and perform matrix subtraction.

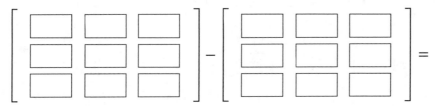

Algebra I Station Activities for Common Core State Standards

Number and Quantity
Set 1: Matrices

As a group, perform matrix subtraction on the following matrices. If matrix subtraction is not possible, explain why. Discuss your answers.

3. $\begin{bmatrix} 3 & -5 \\ 44 & 0 \end{bmatrix} - \begin{bmatrix} 1 & -25 \\ 12 & -18 \end{bmatrix} =$

4. $\begin{bmatrix} 0 & 22 & -11 \\ 15 & 44 & 72 \\ -16 & 8 & -1 \end{bmatrix} - \begin{bmatrix} 0 & 8 & -4 \\ -14 & 7 & 12 \\ 22 & 39 & -21 \end{bmatrix} =$

5. $\begin{bmatrix} 10 & 44 & 29 \end{bmatrix} - \begin{bmatrix} 2 \\ 15 \\ 7 \end{bmatrix} =$

Number and Quantity
Set 1: Matrices

Station 3

You will be given 12 index cards with the following scalars and matrices on them:

Scalars: $-2, -3, 1, 3, 4, 5, 8, 12$

Matrices: $\begin{bmatrix} 10 & -4 & 5 \end{bmatrix}$, $\begin{bmatrix} 12 & 7 \\ 8 & 3 \end{bmatrix}$, $\begin{bmatrix} 10 & 2 & 3 \\ -2 & 4 & 12 \\ 7 & 11 & 3 \end{bmatrix}$, $\begin{bmatrix} -5 & -8 \\ -9 & -3 \end{bmatrix}$

Place the scalars in one pile and the matrices in another pile. Have one student draw a card from each pile. Work together to perform scalar multiplication based on the two cards drawn. Repeat this process two more times and write the scalar, matrix, and your answer in the space below.

1. Draw two cards. Write down the scalar and matrix, then perform scalar multiplication. Show your work.

2. Draw two more cards. Write down the scalar and matrix, then perform scalar multiplication. Show your work.

3. Draw two more cards. Write down the scalar and matrix, then perform scalar multiplication. Show your work.

continued

Number and Quantity
Set 1: Matrices

4. What scalar was used in the following problem? Explain your answer.

$$Scalar \begin{bmatrix} -2 & 10 \\ 3 & 8 \end{bmatrix} = \begin{bmatrix} 10 & -50 \\ -15 & -40 \end{bmatrix}$$

5. What original matrix was used in the following problem? Explain your answer.

$$3 \begin{bmatrix} & \\ & \end{bmatrix} = \begin{bmatrix} -3 & 27 \\ -15 & 12 \end{bmatrix}$$

Algebra I Station Activities for Common Core State Standards

Number and Quantity
Set 1: Matrices

Station 4

Use matrix multiplication to solve the problems. You will be given a number cube to solve problems 2 and 3.

1. As a group, determine how to perform matrix multiplication from the given problem and answer:

$$\begin{bmatrix} 2 & -5 & 9 \\ 10 & 4 & 1 \end{bmatrix} \begin{bmatrix} 7 & 12 \\ 8 & 3 \\ 11 & -1 \end{bmatrix} = \begin{bmatrix} 2(7)+(-5)(8)+(9)(11) & 2(12)+(-5)(3)+(9)(-1) \\ 10(7)+(4)(8)+(1)(11) & 10(12)+(4)(3)+(1)(-1) \end{bmatrix} =$$

$$\begin{bmatrix} 14-40+99 & 24-15-9 \\ 70+32+11 & 120+12-1 \end{bmatrix} = \begin{bmatrix} 73 & 0 \\ 113 & 131 \end{bmatrix}$$

Write your explanation of how to perform matrix multiplication below.

2. Take turns rolling the number cube, for a total of 4 rolls. Each time, record the result in a box in the matrix problem below. When all the boxes in the matrix problem contain a number, perform matrix multiplication and write your answer in the space below.

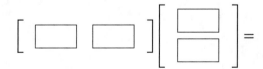

Number and Quantity
Set 1: Matrices

3. Work as a group to roll the number cube and record the result in a box in the matrix problem below. Repeat this process until all the boxes in the matrix problem contain a number. Then perform matrix multiplication and write your answer in the space below.

$$\begin{bmatrix} \square & \square & \square \\ \square & \square & \square \end{bmatrix} \begin{bmatrix} \square & \square \\ \square & \square \\ \square & \square \end{bmatrix} =$$

Number and Quantity

Goal: To provide opportunities for students to develop concepts and skills related to unit
conversion, finding percents, simplifying algebraic ratios, and solving algebraic proportions

Common Core Standards

Number and Quantity: Quantities★

Reason quantitatively and use units to solve problems.

N-Q.1. Use units as a way to understand problems and to guide the solution of multi-step
problems; choose and interpret units consistently in formulas; choose and interpret
the scale and the origin in graphs and data displays.

Algebra: Creating Equations★

Create equations that describe numbers or relationships.

A-CED.1. Create equations and inequalities in one variable and use them to solve problems.

Algebra: Reasoning with Equations and Inequalities

Solve equations and inequalities in one variable.

A-REI.3. Solve linear equations and inequalities in one variable, including equations with
coefficients represented by letters.

Student Activities Overview and Answer Key

Station 1

Students will be given 12 index cards with pairs of equivalent units of measurement written on them.
They will work together to match the cards that are an equivalent unit of measurement. Then they
will perform unit conversion.

Answers

1. 10 mm = 1 cm; 12 in. = 1 ft; 3 ft = 1 yd; 2 pints = 1 quart; 4 quarts = 1 gallon;
   1 ton = 2,000 pounds

2. 8 pints in a gallon; 2 pints = 1 quart and 4 quarts = 1 gallon, so 2(4) = 8 pints

3. 18 inches; 1/2 yard = 1.5 feet and 12 inches = 1 foot, so 12(1.5) = 18 inches

4. 5,000 pounds

5. 850 mm

6. 13.5 feet

7. 3 quarts = 0.75 gallons

8. Answers will vary. Possible answers include: cooking, when modifying recipes for more or fewer people; carpentry, when creating custom-size cabinetry

Station 2

Students will be given a calculator to help them solve the problems. They work as a group to solve real-world applications of unit conversions.

Answers

1. His friend measures temperature in Celsius, and Evan measures it in Fahrenheit. $F = 95°$

2. $P = 36.67$ yards; $P = 1,320$ inches, $A = 77.78$ yds^2; $A = 100,800$ in^2

3.

	Feet	Yards	Meters	Time
Tim	300	100	91.44	12 seconds
Jeremy	400	133.33	121.95	12 seconds
Martin	229.66	76.55	70	12 seconds

Jeremy, Tim, Martin; Tim = 25 feet/sec, Jeremy = 33.33 feet/sec; Martin = 19.14 feet/sec

Station 3

Students will be given a bag containing 24 green marbles and 16 yellow marbles. They will use the marbles to create ratios and percents. They will then solve percent problems.

Answers

1. Answers will vary. Possible answers include: green = 1; yellow = 7; total = 8. Find 1/8 = 0.125 = 12.5%; 12.5% were green. Subtract 12.5% from 100% to get 87.5% or 7/8 = 87.5%; 87.5% were yellow.

2. There are 40 marbles so 24/40 = 60% green marbles; 100% – 60% = 40% or 16/40 = 40%

3. 9 yellow marbles; student drawings should depict 9 yellow marbles and 12 green marbles.

4. 24(1/4) = 6 or 24(0.25) = 6

5. 17(2/1) = 34 or 17(2.0) = 34

6. 10(14) = 140 in^2; increased dimensions by 200% then found the area of the photograph

4. $24(1/4) = 6$ or $24(0.25) = 6$

5. $17(2/1) = 34$ or $17(2.0) = 34$

6. $10(14) = 140$ in^2; increased dimensions by 200% then found the area of the photograph

Station 4

Students will be given 8 large blue algebra tiles and 20 small yellow algebra tiles. Students visually depict ratios and proportions with the algebra tiles. They then solve proportions for a specified variable including a real-world application.

Answers

1. $\dfrac{8 \text{ blue}}{20 \text{ yellow}} = \dfrac{2}{5}$

2. $\dfrac{2 \text{ blue}}{3 \text{ yellow}} = \dfrac{4 \text{ blue}}{6 \text{ yellow}}$

3. $8/20 = x/100$, so $x = 40$ blue

4. $8/20 = x/15$, so $x = 6$ blue

5. $x = 4$

6. $x = 40$

7. $\dfrac{\text{blue}}{\text{yellow}} = \dfrac{6}{10} = \dfrac{3}{5}$

    Let x = number of blue pencils and $24 - x$ = number of yellow pencils.

    $\dfrac{3}{5} = \dfrac{x}{(24 - x)}$, so $x = 9$ blue pencils and $24 - x = 15$ yellow pencils

Materials List/Setup

Station 1   12 index cards with the following written on them:

10 millimeters, 12 inches, 3 feet, 2 pints, 4 quarts, 1 ton, 1 centimeter, 1 foot, 1 yard, 1 quart, 1 gallon, 2,000 pounds

Station 2   calculator

Station 3   24 green marbles; 16 yellow marbles

Station 4   8 large blue algebra tiles; 20 small yellow algebra tiles

Discussion Guide

To support students in reflecting on the activities and to gather some formative information about student learning, use the following prompts to facilitate a class discussion to "debrief" the station activities.

Prompts/Questions

1. How do you perform unit conversion?

2. When would you use unit conversion in the real world?

3. What are two ways to find the percent of a number?

4. What is a ratio?

5. How do you know if two ratios are equivalent?

6. What is a proportion?

7. When would you use ratios and proportions in the real world?

Think, Pair, Share

Have students jot down their own responses to questions, then discuss with a partner (who was not in their station group), and then discuss as a whole class.

Suggested Appropriate Responses

1. Use ratios and proportions to convert units.

2. Answers will vary. Possible answers include: creating scale models of buildings; using the metric system instead of U.S. Customary units; converting Celsius to degrees Fahrenheit and vice versa

3. Multiply the number by a decimal or fraction that represents the percentage.

4. A ratio is a comparison of two numbers by division.

5. Two ratios are equivalent if, when simplified, they are equal.

6. A proportion is when two ratios are set equal to each other.

7. Answers will vary. Possible answers include: enlarging photos; scale models; modifying quantities of ingredients in a recipe

Possible Misunderstandings/Mistakes

- Not keeping track of units and using incorrect unit conversions
- Not recognizing that terms must have the same units in order to compare them
- Setting up proportions with one of the ratios written with the incorrect numbers in the numerator and denominator
- Not recognizing simplified forms of ratios in order to find equivalent ratios

Number and Quantity
Set 2: Ratios and Proportions

Station 1

You will be given 12 index cards with the following written on them:

>10 millimeters, 12 inches, 3 feet, 2 pints, 4 quarts, 1 ton, 1 centimeter, 1 foot, 1 yard,
>1 quart, 1 gallon, 2,000 pounds

Shuffle the index cards and deal a card to each student in your group until all the cards are gone. As a group, show your cards to each other and match the cards that are an equivalent unit of measurement.

1. Write your answers on the lines below. The first match is shown:

    _____10 mm = 1 cm_____   _____

    _____   _____

    _____   _____

2. Find the number of pints in a gallon. Explain how you can use your answers in problem 1 to find the number of pints in a gallon.

3. Find the number of inches in half of a yard. Explain how you can use your answers in problem 1 to find the number of inches in half of a yard.

continued

Number and Quantity
Set 2: Ratios and Proportions

Perform the following unit conversions by filling in the blanks.

4. 2.5 tons = _____ pounds

5. 85 cm = _____ mm

6. 4.5 yd = _____ ft

7. 6 pints = _____ quarts = _____ gallons

8. When would you use unit conversions in the real world?

Number and Quantity
Set 2: Ratios and Proportions

Station 2

You will be given a calculator to help you solve the problems. Work as a group to solve these real-world applications of unit conversions.

1. Evan has a friend in England. His friend said the temperature was very hot at 35°. Evan thought he heard his friend incorrectly since 35° is cold. What caused his misunderstanding? (*Hint*: $C = (F - 32)\dfrac{5}{9}$)

    Find the equivalent temperature in the United States that would make the claim of Evan's friend valid. Write your answer in the space below.

2. Anna is going to build a patio. She wants the patio to be 20 feet by 35 feet. What is the perimeter of the patio in yards?

    What is the perimeter of the patio in inches?

    What is the area of the patio in yards?

continued

What is the area of the patio in inches?

3. Tim claims he can run the 100-yard dash in 12 seconds. Jeremy claims he can run 400 feet in 12 seconds. Martin claims he can run 70 meters in 12 seconds. (*Hint*: 1 yard = 0.9144 meters and 1 yard = 3 feet.)

Fill in the table below to create equivalent units of measure.

	Feet	**Yards**	**Meters**	**Time (seconds)**
Tim				
Jeremy				
Martin				

List the three boys in order of fastest to slowest:

How fast did each boy run in feet/second?

Number and Quantity
Set 2: Ratios and Proportions

Station 3

You will be given a bag containing 24 green marbles and 16 yellow marbles. You will use the marbles to create ratios and percents. You will then solve percent problems. Work together as a group to solve the following problems.

1. Shake the bag of green and yellow marbles so that the colors are mixed. Have each student select 2 marbles from the bag without looking. Group all your marbles together by color.

    How many green marbles did you draw? _____

    How many yellow marbles did you draw? _____

    What was the total number of marbles drawn? _____

    How can you determine the percentage of marbles that were green?

    Find the percentage of marbles you drew that were green.

    Name two ways you can find the percentage of marbles you drew that were yellow.

    Find the percentage of marbles you drew that were yellow.

2. Take all the marbles out of the bag. How can you determine what percentage of all the marbles are green?

    How can you determine what percentage of all the marbles are yellow?

continued

Number and Quantity
Set 2: Ratios and Proportions

3. Place 12 green marbles on the table. How many yellow marbles do you need to have 75% as many yellow marbles on the table?

   Draw a picture of the number of green marbles and yellow marbles you have placed on the table.

4. Use equations to show two ways you can find 25% of 24.

5. Use equations to show two ways you can find 200% of 17.

6. Real-world application: Bryan is a photographer. He has a 5 in. by 7 in. photo that he wants to enlarge by 200%. What is the area of the new photo? Explain your answer in the space below.

Number and Quantity
Set 2: Ratios and Proportions

Station 4

You will be given 8 large blue algebra tiles and 20 small yellow algebra tiles. Work as a group to arrange the algebra tiles so they visually depict the ratio of blue to yellow algebra tiles.

1. What is this ratio? _____

Rearrange the tiles to visually depict the following ratios:

$$\frac{2 \text{ blue}}{3 \text{ yellow}} \qquad \frac{1 \text{ blue}}{10 \text{ yellow}} \qquad \frac{4 \text{ blue}}{6 \text{ yellow}} \qquad \frac{1 \text{ blue}}{1 \text{ yellow}}$$

2. Which ratios are equivalent ratios? Explain your answer.

3. Keeping the same ratio of yellow to blue tiles, if there were 100 yellow algebra tiles, how many blue algebra tiles would there be? Use a proportion to solve this problem. Show your work in the space below. (*Hint*: A proportion is two ratios that are equal to each other.)

4. Keeping the same ratio of yellow to blue tiles, if there were 15 yellow algebra tiles, how many blue algebra tiles would there be? Use a proportion to solve this problem. Show your work in the space below.

continued

Number and Quantity

Set 2: Ratios and Proportions

Work together to solve the following proportions for the variable.

5. $\dfrac{2}{7} = \dfrac{x}{14}$; $x =$

6. $\dfrac{8}{x} = \dfrac{2}{10}$; $x =$

Use the following information to answer problem 7:

Allison has 6 blue pencils and 10 yellow pencils. Sadie has 24 pencils that are either blue or yellow. The ratio of blue pencils to yellow pencils is the same for both Allison and Sadie.

7. How many blue pencils and yellow pencils does Sadie have? Show your work in the space below by setting up a proportion using a variable, x.

Seeing Structure in Expressions

Set 1: Simplifying Radical Expressions with Variables

Goal: To provide opportunities for students to develop concepts and skills related to simplifying radical expressions with variables

Common Core Standards

Algebra: Seeing Structure in Expressions

Interpret the structure of expressions.

    A-SSE.2.   Use the structure of an expression to identify ways to rewrite it.

Write expressions in equivalent forms to solve problems.

    A-SSE.3.   Choose and produce an equivalent form of an expression to reveal and explain properties of the quantity represented by the expression.★

Student Activities Overview and Answer Key

Station 1

Students will be given a number cube. Students will roll the number cube in order to populate radical expressions with variables. Then they will simplify these radical expressions by using the law of exponents.

Answers

1. 1/2

2. Answers will vary. Possible answer: $\left(x^2 y^4 z^5 \right)^{\frac{1}{2}}$

3. Answers will vary. Possible answer:

$$\left(x^2 y^4 z^5 \right)^{\frac{1}{2}} = xy^2 z^{\frac{5}{2}}$$

$$= xy^2 \left(z^{\frac{4}{2}} \bullet z^{\frac{1}{2}} \right) = xy^2 \left(z^2 \bullet z^{\frac{1}{2}} \right)$$

$$= xy^2 z^2 \sqrt{z}$$

4. We wrote the expression with a rational exponent, then used the law of exponents to simplify the expression.

5. Answers will vary. Possible answer: $\left(x^3 y z^6 \right)^{\frac{1}{2}}$

6. Answers will vary. Possible answer:

$$\left(x^3 y z^6\right)^{\frac{1}{2}} = x^{\frac{3}{2}} y^{\frac{1}{2}} z^3$$

$$= \left(x^{\frac{2}{2}} \bullet x^{\frac{1}{2}}\right) y^{\frac{1}{2}} z^3 = \left(x^1 \bullet x^{\frac{1}{2}}\right) y^{\frac{1}{2}} z^3$$

$$= x\sqrt{x} \bullet \sqrt{y} \bullet z^3$$

$$= xz^3 \sqrt{xy}$$

7. $4x^4 y$

8. $d^{10} f^5 g \sqrt{fg}$

9. $2rt^5 \sqrt{2r}$

10. $\dfrac{a^2 b^6 c^4 \sqrt{3ac}}{6}$

Station 2

Students will be given 25 blue algebra tiles, 25 green algebra tiles, and 20 red algebra tiles. Students will use the algebra tiles to modify and simplify radical expressions with variables. They will then describe how to use the law of exponents to simplify radical expressions.

Answers

1a. 4

1b. $\left(a^8\right)^{\frac{1}{2}} \left(b^4\right)^{\frac{1}{2}}$

1c. $4a^4 b^2$

2a. $2\sqrt{3}$

2b. $\left(a^9\right)^{\frac{1}{2}} \left(b^5\right)^{\frac{1}{2}} = \left(a^{\frac{9}{2}}\right)\left(b^{\frac{5}{2}}\right)$

$$= a^4 b^2 \sqrt{ab}$$

2c. $2a^4 b^2 \sqrt{3ab}$

3. $3a^3 b^6 \sqrt{2a}$

Station 3

Students will be given six index cards with the following written on them:

$$x^3 \bullet x^{\frac{1}{2}} \; ; \; x^{\frac{6}{2}} \bullet x^{\frac{1}{2}} \; ; \; x^{\frac{7}{2}} \; ; \; \sqrt{x^7} \; ; \; (x^7)^{\frac{1}{2}} \; ; \; x^3 \sqrt{x}$$

Students work together to arrange the index cards to model how to change a term written in exponents to a term written as a radical.

Answers

1. $\sqrt{x^7}$

2. $(x^7)^{\frac{1}{2}}$

3. $x^{\frac{7}{2}}$

4. $x^{\frac{6}{2}} \bullet x^{\frac{1}{2}}$

5. $x^3 \bullet x^{\frac{1}{2}}$

6. $x^3 \sqrt{x}$

7. Answers will vary.

8. $yz^2 \sqrt{yz}$

Station 4

Students will be given eight index cards with the following radicals written on them:

$$\sqrt{8x^4 y} \; ; \; \sqrt{16x^3 y^6} \; ; \; \sqrt{27x^9 y^5} \; ; \; \sqrt{\frac{27x^3 y^7}{3}} \; ; \; 2x^2 \sqrt{2y} \; ; \; 4xy^3 \sqrt{x} \; ; \; 3x^4 y^2 \sqrt{3xy} \; ; \; 3xy^3 \sqrt{xy}$$

Students will work as a group to match radicals with their simplified counterparts and then explain how to simplify the radical showing the steps towards simplification.

Answers

Non-simplified radical	Simplified radical	Steps to simplify
$\sqrt{8x^4 y}$	$2x^2 \sqrt{2y}$	$\sqrt{8x^4 y} = \sqrt{4 \bullet 2 \left(x^2\right)^2 y}$ $= 2x^2 \sqrt{2y}$
$\sqrt{16x^3 y^6}$	$4xy^3 \sqrt{x}$	$\sqrt{16x^3 y^6} = \sqrt{4^2 x^2 \bullet x \bullet \left(y^3\right)^2}$ $= 4xy^3 \sqrt{x}$
$\sqrt{27x^9 y^5}$	$3x^4 y^2 \sqrt{3xy}$	$\sqrt{27x^9 y^5}$ $= \sqrt{3^2 \bullet 3 \bullet \left(x^2\right)^4 \bullet x \bullet \left(y^2\right)^2 \bullet y}$ $= 3x^4 y^2 \sqrt{3xy}$
$\sqrt{\dfrac{27x^3 y^7}{3}}$	$3xy^2 \sqrt{xy}$	$\sqrt{\dfrac{27x^3 y^7}{3}} = \sqrt{9x^3 y^7}$ $= \sqrt{3^2 x^2 \bullet x \bullet \left(y^2\right)^3 \bullet y}$ $= 3xy^2 \sqrt{xy}$

Materials List/Setup

Station 1   number cube

Station 2   25 blue algebra tiles; 25 green algebra tiles; 20 red algebra tiles

Station 3   six index cards with the following written on them:

$$x^3 \bullet x^{\frac{1}{2}} ; \ x^{\frac{6}{2}} \bullet x^{\frac{1}{2}} ; \ x^{\frac{7}{2}} ; \ \sqrt{x^7} ; \ (x^7)^{\frac{1}{2}} ; \ x^3 \sqrt{x}$$

Station 4   eight index cards with the following radicals written on them:

$$\sqrt{8x^4 y} ; \ \sqrt{16x^3 y^6} ; \ \sqrt{27x^9 y^5} ; \ \sqrt{\dfrac{27x^3 y^7}{3}} ; \ 2x^2 \sqrt{2y} ; \ 4xy^3 \sqrt{x} ; \ 3x^4 y^2 \sqrt{3xy} ; \ 3xy^3 \sqrt{xy}$$

Discussion Guide

To support students in reflecting on the activities and to gather some formative information about student learning, use the following prompts to facilitate a class discussion to "debrief" the station activities.

Prompts/Questions

1. How do you simplify radical expressions?

2. Why does writing the radical in exponential form help you simplify the radical?

3. How can you change a radical expression written as fractional exponents into a radical expression?

4. Do you have to take the square root of each base separately underneath a square root sign? Why or why not?

5. When would you use radicals in the real world?

Think, Pair, Share

Have students jot down their own responses to questions, then discuss with a partner (who was not in their station group), and then discuss as a whole class.

Suggested Appropriate Responses

1. Write the radical as a fractional exponent. Then use the law of exponents to simplify the radical expression.

2. You can multiply the fractional exponent by the exponent of each base in the expression.

3. Use the denominator of the fractional exponent as the root. Then simplify out appropriate terms based on their exponent's relation to the root.

4. Yes, you have to apply the square root to each base underneath the square root sign because the bases show the square of a number or base.

5. Answers may vary. Sample answer: You use radicals when using the Pythagorean theorem in geometry and construction.

Possible Misunderstandings/Mistakes

- Not applying the square root to each base of the radicand including the constant

- Not understanding how to write a radical as a fractional exponent

- Incorrectly using the law of exponents to simplify the expressions

Seeing Structure in Expressions
Set 1: Simplifying Radical Expressions with Variables

Station 1

You will be given a number cube. As a group, roll the number cube. Write the result in the first box. Repeat this process until each box contains a number.

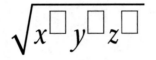

$$\sqrt{x^\square y^\square z^\square}$$

1. How can you write $\sqrt{}$ as an exponent?

   _____

2. Rewrite the expression you created by rolling the number cube, raised to the exponent from problem 1.

   _____

3. Simplify this expression. Show your work.

4. What strategy did you use to simplify this expression?

As a group, roll the number cube. Write the result in the first box. Repeat this process until each box contains a number.

$$\sqrt{x^\square y^\square z^\square}$$

5. Rewrite the expression above raised to an exponent.

   _____

continued

Seeing Structure in Expressions
Set 1: Simplifying Radical Expressions with Variables

6. Simplify this expression. Show your work.

Work together to simplify each expression below.

7. $\sqrt{16x^8 y^2} = $ _____

8. $\sqrt{d^{20} f^{11} g^3} = $ _____

9. $\sqrt{8r^3 t^{10}} = $ _____

10. $\sqrt{\dfrac{a^5 b^{12} c^9}{12}} = $ _____

Seeing Structure in Expressions
Set 1: Simplifying Radical Expressions with Variables

Station 2

You will be given 25 blue algebra tiles, 25 green algebra tiles, and 20 red algebra tiles. Work as a group to model each problem by placing the tiles in the boxes. Then move the tiles around to simplify each expression.

- Use the blue tiles to represent the constant.
- Use the green tiles to represent the exponent of a.
- Use the red tiles to represent the exponent of b.

1. Simplify $\sqrt{16a^8b^4}$.

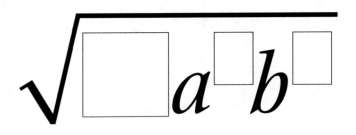

a. What is the square root of the constant?

_____

b. How can you use the law of exponents to find $\sqrt{a^8b^4}$?

c. How can you write $\sqrt{16a^8b^4}$ in simplified form?

_____

continued

Algebra I Station Activities for Common Core State Standards

Seeing Structure in Expressions
Set 1: Simplifying Radical Expressions with Variables

2. Simplify $\sqrt{12a^9 b^5}$.

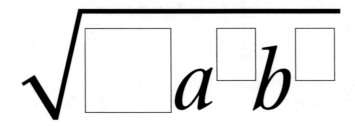

a. What is the square root of the constant?

_____

b. How can you use the law of exponents to find $\sqrt{a^9 b^5}$?

c. How can you write $\sqrt{12a^9 b^5}$ in simplified form?

_____

3. Simplify $\sqrt{18a^7 b^{12}}$. Show your work.

Seeing Structure in Expressions
Set 1: Simplifying Radical Expressions with Variables

Station 3

You will be given six index cards with the following written on them:

$$x^3 \bullet x^{\frac{1}{2}} \; ; \; x^{\frac{6}{2}} \bullet x^{\frac{1}{2}} \; ; \; x^{\frac{7}{2}} \; ; \; \sqrt{x^7} \; ; \; (x^7)^{\frac{1}{2}} \; ; \; x^3 \sqrt{x}$$

Shuffle the cards. Work together to put the cards in an order that changes $x^{\frac{7}{2}}$ into a simplified radical expression. Write the order of the index cards on the lines below.

1. _____

2. _____

3. _____

4. _____

5. _____

6. _____

7. What strategy did you use to arrange the index cards?

8. As a group, simplify $\sqrt{y^3 z^5}$. Show your work.

Seeing Structure in Expressions
Set 1: Simplifying Radical Expressions with Variables

Station 4

At this station, you will find eight index cards with the following expressions:

$$\sqrt{8x^4 y} \; ; \; \sqrt{16x^3 y^6} \; ; \; \sqrt{27x^9 y^5} \; ; \; \sqrt{\frac{27x^3 y^7}{3}} \; ; \; 2x^2\sqrt{2y} \; ; \; 4xy^3\sqrt{x} \; ; \; 3x^4 y^2 \sqrt{3xy} \; ; \; 3xy^3\sqrt{xy}$$

Work with your group to match the non-simplified expression with its simplified version. Write the matching pairs you find, then write the steps needed to get from the non-simplified version to the simplified version.

Non-simplified radical	Simplified radical	Steps to simplify

Seeing Structure in Expressions

Set 2: Operations with Radicals and Variables

Goal: To provide opportunities for students to develop concepts and skills related to adding, subtracting, multiplying, and dividing radical expressions. Students will also solve radical equations.

Common Core Standards

Algebra: Seeing Structure in Expressions

Interpret the structure of expressions.

    A-SSE.2.   Use the structure of an expression to identify ways to rewrite it.

Write expressions in equivalent forms to solve problems.

    A-SSE.3.   Choose and produce an equivalent form of an expression to reveal and explain properties of the quantity represented by the expression.★

Student Activities Overview and Answer Key

Station 1

Students will be given eight index cards with the following written on them:

$$-2\sqrt{4x}\;;\;\sqrt{49x}\;;\;-3\sqrt{8x}\;;\;-2\sqrt{18x}\;;\;\sqrt{12x}\;;\;-\sqrt{27x}\;;\;-\sqrt{20x}\;;\;\sqrt{45x}$$

along with four index cards with the following written on them:

$$\Box\sqrt{x}\;;\;\Box\sqrt{2x}\;;\;\Box\sqrt{3x}\;;\;\Box\sqrt{5x}$$

Students will work together to find the radicals that simplify to the four radicands given on the cards and then they will combine the terms.

Answers

1. $\Box\sqrt{x}:-2\sqrt{4x}+\sqrt{49x}=-4\sqrt{x}+7\sqrt{x}=3\sqrt{x}$
2. $\Box\sqrt{2x}:-3\sqrt{8x}-2\sqrt{18x}=-6\sqrt{2x}-6\sqrt{2x}=-12\sqrt{2x}$
3. $\Box\sqrt{3x}:\sqrt{12x}-\sqrt{27x}=2\sqrt{3x}-3\sqrt{3x}=-\sqrt{3x}$
4. $\Box\sqrt{5x}:-\sqrt{20x}+\sqrt{45x}=-2\sqrt{5x}+3\sqrt{5x}=\sqrt{5x}$
5. $3xy\sqrt{3}+4xy\sqrt{3}-20xy\sqrt{3}=-13xy\sqrt{3}$

Station 2

Students will be given three index cards with the following terms written on them:

$$4\sqrt{2xy} \; ; \; 3\sqrt{6xz} \; ; \; 2\sqrt{10yz}$$

They will also be given three index cards with the following answers written on them:

$$24x\sqrt{3yz} \; ; \; 16y\sqrt{5xz} \; ; \; 12z\sqrt{15xy}$$

Students will match the appropriate "answer" card with the two "terms" cards that when multiplied together yield that answer. They will write the steps they used in multiplying the radicals and simplifying them.

Answers

1. $4\sqrt{2xy} \bullet 3\sqrt{6xy} = 24x\sqrt{3yz}$

2. $4\sqrt{2xy} \bullet 2\sqrt{10yz} = 16y\sqrt{5xz}$

3. $3\sqrt{6xy} \bullet 2\sqrt{10yz} = 12z\sqrt{15xy}$

4. Answers will vary. Possible answer: We multiplied the numbers outside the radicals first, then multiplied the radicands and simplified them.

5. The numbers and variables underneath the radical sign do not have to be the same when multiplying. The reason is because you use the product rule of the law of exponents to deal with each base under the radical.

6. $4xy^2\sqrt{3x}$

7. $6xy^4\sqrt{5x}$

8. These radicals are not simplified.

9. Answers will vary. Sample answer: Simplify the radicals first, then multiply and simplify again.

Station 3

Students will be given 20 blue algebra tiles and 20 red algebra tiles. Students will use the algebra tiles to model and solve division of radicals with variables. They will discover why the index of the radical is important, including when to use the product rule. Then they will divide radicals with variables.

Answers

1. 2

2. 4

3. Quotient rule: when dividing exponents with the same base, subtract the powers.

4. $x^5 \div x^3 = x^{5-3} = x^2$; $2\sqrt{4x^2}$

5. 2

6. Find perfect squares under the radical sign.

7. $4x$

8. Divide the numbers in front of the radical signs. Divide the numbers and/or variables under the radical sign using the quotient rule. Then simplify your answer.

9. $75a\sqrt{30b^{15}c} \div 25a\sqrt{6b^2c} = 3\sqrt{5b^{13}} = 3b^6\sqrt{5b}$

10. $\dfrac{2}{3}x^8y\sqrt{48x^{12}z} \div \dfrac{4}{5}x^3\sqrt{2x^7} = \dfrac{5}{6}x^5y\sqrt{24x^5z} = \dfrac{5}{3}x^7y\sqrt{6xz}$

Station 4

Students will be given four index cards with the following radical equations written on them:

$$\sqrt{6x} - 4 = 2;\ \sqrt{3x+4} = 5;\ 2\sqrt{x} - 5 = 3;\ 3\sqrt{x+1} = 21$$

Then they will be given cards with steps written on them. Their task is to put the steps in order underneath the equation, ending with the card showing the solution.

$$\sqrt{6x} = 6;\ 6x = 36;\ x = 6;\ 3x + 4 = 25;\ 3x = 21;\ x = 7;\ \sqrt{x} = 4;\ x = 16;\ \sqrt{x+1} = 7;\ x = 48$$

Students will work together to match the radical equations with the appropriate step in order to solve the equation. Then they will solve each multistep radical equation, showing all the steps involved. Then they will solve a radical equation that has multiple steps.

Answers

1.

Equation	Steps	Final answer
$\sqrt{6x} - 4 = 2$	$\sqrt{6x} - 4 = 2$ $\sqrt{6x} = 6$ $\left(\sqrt{6x}\right)^2 = \left(6\right)^2$ $6x = 36$ $x = 6$	$x = 6$
$\sqrt{3x + 4} = 5$	$\sqrt{3x + 4} = 5$ $\left(\sqrt{3x + 4}\right)^2 = \left(5\right)^2$ $3x + 4 = 25$ $3x = 21$ $x = 7$	$x = 7$
$2\sqrt{x} - 5 = 3$	$2\sqrt{x} - 5 = 3$ $2\sqrt{x} = 8$ $\sqrt{x} = 4$ $\left(\sqrt{x}\right)^2 = \left(4\right)^2$ $x = 16$	$x = 16$
$3\sqrt{x + 1} = 21$	$3\sqrt{x + 1} = 21$ $\sqrt{x + 1} = 7$ $\left(\sqrt{x + 1}\right)^2 = \left(7\right)^2$ $x + 1 = 49$ $x = 48$	$x = 48$

Algebra I Station Activities for Common Core State Standards

2. $5 + 4\sqrt{8x + 4} = 29$

    $4\sqrt{8x + 4} = 24$

    $\sqrt{8x + 4} = 6$

    $8x + 4 = 36$

    $8x = 32$

    $x = 4$

Materials List/Setup

Station 1 eight index cards with the following written on them:

$-2\sqrt{4x}$; $\sqrt{49x}$; $-3\sqrt{8x}$; $-2\sqrt{18x}$; $\sqrt{12x}$; $-\sqrt{27x}$; $-\sqrt{20x}$; $\sqrt{45x}$

also, four index cards with the following written on them:

$\square\sqrt{x}$; $\square\sqrt{2x}$; $\square\sqrt{3x}$; $\square\sqrt{5x}$

Station 2 six index cards with the following terms written on them:

$4\sqrt{2xy}$; $3\sqrt{6xy}$; $2\sqrt{10yz}$

also, three index cards with the following answers written on them:

$24x\sqrt{3yz}$; $16y\sqrt{5xz}$; $12z\sqrt{15xy}$

Station 3 20 blue algebra tiles; 20 red algebra tiles

Station 4 four index cards with the following radical equations written on them:

$\sqrt{6x} - 4 = 2$; $\sqrt{3x + 4} = 5$; $2\sqrt{x} - 5 = 3$; $3\sqrt{x + 1} = 21$

ten index cards with the following:

$\sqrt{6x} = 6$; $6x = 36$; $x = 6$; $3x + 4 = 25$; $3x = 21$; $x = 7$; $\sqrt{x} = 4$; $x = 16$;

$\sqrt{x + 1} = 7$; $x = 48$

Discussion Guide

To support students in reflecting on the activities and to gather some formative information about student learning, use the following prompts to facilitate a class discussion to "debrief" the station activities.

Prompts/Questions

1. How do you add and subtract radicals with variables?

2. What has to be equal in order to add and subtract radicals?

3. How do you multiply radicals with variables?

4. How do you divide radicals with variables?

5. How do you solve radical equations?

Think, Pair, Share

Have students jot down their own responses to questions, then discuss with a partner (who was not in their station group), and then discuss as a whole class.

Suggested Appropriate Responses

1. You add and subtract the numbers in front of the radicand.

2. the radicands

3. Multiply the numbers in front of the radical. Multiply the numbers within the radical. Use the product rule on the exponents of the variables.

4. Divide the numbers in front of the radical. Divide the numbers within the radical. Use the quotient rule on the exponents of the variables.

5. Isolate the variable. Square each side to remove the radical. Then solve for the variable.

Possible Misunderstandings/Mistakes

- Adding and subtracting radicals that do not have the same radicand
- Forgetting to also multiply the numbers and variables in the radicand when multiplying radicals
- Forgetting to also divide the numbers and variables in the radicand when dividing radicals
- Not writing the radical in the simplest form
- Forgetting to apply the square to both sides of the equation when solving for the variable

Seeing Structure in Expressions
Set 2: Operations with Radicals and Variables

Station 1

At this station, you will find eight index cards with the following written on them:

$$-2\sqrt{4x}\;;\; \sqrt{49x}\;;\; -3\sqrt{8x}\;;\; -2\sqrt{18x}\;;\; \sqrt{12x}\;;\; -\sqrt{27x}\;;\; -\sqrt{20x}\;;\; \sqrt{45x}$$

As a group, shuffle the cards. Each person draws one card at a time, until all the cards have been dealt.

You will also find four index cards with the following written on them:

$$\boxed{}\sqrt{x}\;;\; \boxed{}\sqrt{2x}\;;\; \boxed{}\sqrt{3x}\;;\; \boxed{}\sqrt{5x}$$

Place these cards face down in a pile.

One person draws a card from this pile and places it face up on the desk. Everyone should look at the cards in their hand. Two players will have cards in their hands that simplify with the same radicand as in the face-up card. Those two students will put the cards with the matching radicands on the table for everyone to see. Then they will simplify and combine the terms.

Write the results below for the given radicand.

1. Radicand: $\boxed{}\sqrt{x}$

   Cards with matching radicands: _____

   Simplify and combine. Show your work.

2. Radicand: $\boxed{}\sqrt{2x}$

   Cards with matching radicands: _____

   Simplify and combine. Show your work.

© 2011 Walch Education
Algebra I Station Activities for Common Core State Standards

Seeing Structure in Expressions
Set 2: Operations with Radicals and Variables

3. Radicand: $\boxed{}\sqrt{3x}$

    Cards with matching radicands: _____

    Simplify and combine. Show your work.

4. Radicand: $\boxed{}\sqrt{5x}$

    Cards with matching radicands: _____

    Simplify and combine. Show your work.

5. Work together to simplify the equation below. Show your work.

    $$x\sqrt{27y^2} + 2y\sqrt{12x^2} - 5xy\sqrt{48}$$

Seeing Structure in Expressions
Set 2: Operations with Radicals and Variables

Station 2

You will be given six index cards with the following terms written on them:

$$4\sqrt{2xy} \; ; \; 3\sqrt{6xz} \; ; \; 2\sqrt{10yz}$$

You will also be given three index cards with the following answers written on them:

$$24x\sqrt{3yz} \; ; \; 16y\sqrt{5xz} \; ; \; 12z\sqrt{15xy}$$

Work together to match the two "terms" cards that could be multiplied together to form the "answer" card. Write your answer below in the format *term card • term card = answer card*.

1. _____ • _____ = _____

    Below, write the steps you used to multiply and simplify the problem.

2. _____ • _____ = _____

    Below, write the steps you used to multiply and simplify the problem.

3. _____ • _____ = _____

    Below, write the steps you used to multiply and simplify the problem.

continued

4. What strategy did you use to match the answers with the "terms cards"?

5. When multiplying radicals does the radicand, the quantity under the radical sign, have to be equal? Why or why not?

Work together to multiply the following radical expressions. Make sure your answer is simplified.

6. $\sqrt{24x^2 y} \bullet \sqrt{2xy^3}$ = _____

7. $\sqrt{60xy^5} \bullet \sqrt{3x^2 y^3}$ = _____

8. What is different about these radicals compared to the radicals you worked with in the first set of multiplication problems?

9. Did your strategy in multiplying these radicals differ from the strategy you used above? If so, explain.

Seeing Structure in Expressions
Set 2: Operations with Radicals and Variables

Station 3

You will be given 20 blue algebra tiles and 20 red algebra tiles. Work together to model the following problem.

- Use blue algebra tiles to represent the number in front of the radical.
- Use red algebra tiles to represent the number underneath the radical.

$$10\sqrt{12x^5} \div 5\sqrt{3x^3}$$

1. Divide the blue algebra tiles. Write the answer in the dashed box below.

$$\boxed{}\ \sqrt{\boxed{}}\ \vdots\vdots$$

2. Divide the red algebra tiles. Write the answer in the solid box under problem 1.

3. What law of exponent rule can you use to determine $x^5 \div x^3$?

4. What is $x^5 \div x^3$? _____

Write this answer in the dotted box under problem 1.

5. What number is understood to be in front of the $\sqrt{}$ sign? _____

6. What does that number mean? _____

7. Simplify the radical. _____

8. Based on your answers in problems 1–7, explain how to divide radicals with variables.

continued

Seeing Structure in Expressions
Set 2: Operations with Radicals and Variables

Work as a group to divide the radicals in each problem below.

9. $75a\sqrt{30b^{15}c} \div 25a\sqrt{6b^2c}$

10. $\dfrac{2}{3}x^8 y\sqrt{48x^{12}z} \div \dfrac{4}{5}x^3\sqrt{2x^7}$

Seeing Structure in Expressions
Set 2: Operations with Radicals and Variables

Station 4

You will be given four index cards with the following radical equations written on them:

$$\sqrt{6x} - 4 = 2; \ \sqrt{3x+4} = 5; \ 2\sqrt{x} - 5 = 3; \ 3\sqrt{x+1} = 21$$

You will be given ten index cards with "steps" written on them:

$$\sqrt{6x} = 6; \ 6x = 36; \ x = 6; \ 3x + 4 = 25; \ 3x = 21; \ x = 7; \ \sqrt{x} = 4; \ x = 16; \ \sqrt{x+1} = 7; \ x = 48$$

Work together to solve each radical equation using the order of operations and rules for radicals.

1. For each equation, write the steps to solve the equation and your final answer. Show all the work that is needed to arrive at the next step card in the table.

Equation	Steps	Final answer
$\sqrt{6x} - 4 = 2$		
$\sqrt{3x+4} = 5$		
$2\sqrt{x} - 5 = 3$		
$3\sqrt{x+1} = 21$		

continued

2. Solve $5 + 4\sqrt{8x + 4} = 29$. Show your work.

Seeing Structure in Expressions

Set 3: Factoring Polynomials

Goal: To provide opportunities for students to develop concepts and skills related to factoring polynomials

Common Core Standards

Algebra: Seeing Structure in Expressions

Interpret the structure of expressions.

  A-SSE.2.　Use the structure of an expression to identify ways to rewrite it.

Write expressions in equivalent forms to solve problems.

  A-SSE.3.　Choose and produce an equivalent form of an expression to reveal and explain properties of the quantity represented by the expression.★

  　　a. Factor a quadratic expression to reveal the zeros of the function it defines.

  　　b. Complete the square in a quadratic expression to reveal the maximum or minimum value of the function it defines.

  　　c. Use the properties of exponents to transform expressions for exponential functions.

Algebra: Arithmetic with Polynomials and Rational Expressions

Use polynomial identities to solve problems.

  A-APR.4.　Prove polynomial identities and use them to describe numerical relationships.

Student Activities Overview and Answer Key

Station 1

Students will be given a number cube. Students will use the number cube to populate the exponents of terms and expressions. They will find the greatest common factor of terms and expressions. Then they will factor the expression using the greatest common factor.

Answers

1. Answers will vary. Possible answer: x^3, x^6, x^4; x^3

2. 1; x^3; x

3. greatest common factor

4. Answers will vary. Possible answer: $4x^3 - 6x^2 + 4x^4$; 2; x^2; $2x^2$; $2x - 3 + 2x^2$

5. Answers will vary. Possible answer: $-5x^2y + x^3 - 10x^5y^4$; 1; x^2; x^2; $-5y + x - 10x^3y^4$

6. Answers will vary. Possible answer: $6x^3yz + 2s + 4x^2y^5$; 2; no common variables; 2; $3x^3yz + s + 2x^2y^5$

7. No, because there was no variable that all three terms had in common.

8. $4x^2y^4z^3(3x^2y + 14z^3 - 6xy^3z^5)$

9. $3c^2(9a^2b^3 - 4ac - 3b^2c^3)$

10. $-9s^2t(4r^2st + 2rs + 3r^2s^2t^4 + 1)$

Station 2

Students will be given eight blank index cards, plus ten index cards with the following written on them:

$$3x; x; +1; +2; +4; +8; -1; -2; -4; -8$$

Students will work together to arrange the cards to factor a trinomial. Then they will create the possible factors of a trinomial and factor the trinomial. Students factor trinomials with a leading coefficient other than 1.

Answers

1. $(3x + 2)(x + 4)$

2. Answers will vary. Possible answer: We used the distribution method to check factors.

3. The factors are $3x$ and x because 3 is a prime number.

4. x, $2x$, $3x$, and $6x$

5. $-5, -1, 1, 5$

6. $(2x - 1)(3x + 5)$

7. Use the distribution method to double-check answers.

Station 3

Students will be given five index cards with the following expressions written on them:

$$x^2 + 8x + 12; \ x^2 - 8x + 15; \ x^2 + 2x - 80; \ x^2 + x - 12; \ x^2 - x - 12$$

They will also receive five index cards with the following factors written on them:

$$(x - 3)(x + 4); \ (x + 10)(x - 8); \ (x + 2)(x + 6); \ (x + 3)(x - 4); \ (x - 3)(x - 5)$$

Students will work together to match each expression with the appropriate factors. Then students will factor trinomials with a leading coefficient of 1. They will explain how to double-check their answers and why factoring out the greatest common factor first is important.

Answers

1. $x^2 + 8x + 12$ and $(x + 2)(x + 6)$

2. $x^2 - 8x + 15$ and $(x - 3)(x - 5)$

3. $x^2 + 2x - 80$ and $(x + 10)(x - 8)$

4. $x^2 + x - 12$ and $(x - 3)(x + 4)$

5. $x^2 - x - 12$ and $(x + 3)(x - 4)$

6. Answers will vary.

7. Use the distribution method to multiply the binomials. This should yield the original trinomial.

8. 1

9. x and x

10. 6 and –2; ($\boxed{x}$ + $\boxed{6}$)($\boxed{x}$ + ($\boxed{-2}$)); $(x + 6)(x - 2)$

11. 2

12. $x^2 + 4x - 5$

13. x and x

14. –5; 4; 5 and –1; ($\boxed{x}$ + $\boxed{5}$)($\boxed{x}$ – $\boxed{1}$); $2(x + 5)(x - 1)$

15. Answers will vary. Possible answer: It is easier to factor smaller numbers.

Station 4

Students will be given a number cube. Students will use the number cube to populate binomial expressions. They will multiply the binomial expressions using the distribution method. Then they will factor the polynomial they created. They will relate the distribution method to factoring. They will factor the difference of squares and perfect square trinomials.

Answers

1. Answers will vary. Possible answer: $(x + 2)(x - 2) = x^2 - 4$

2. 2

3. It cancels out.

4. $(2x + 3)(2x - 3)$; Find the square root of the first term and the second terms. Write the factors in $(a + b)(a - b)$ form.

5. $(7x^3 + 6)(7x^3 - 6)$

6. Answers will vary. Possible answer: $(2x + 3)(2x + 3) = 4x^2 + 12x + 9$

7. 3

8. $(4x + 3)(4x + 3)$; Find the square root of the first term and the third term. Write the factors in $(a + b)(a + b)$ form.

9. $(2x^4 + 5)(2x^4 + 5)$

10. Answers will vary. Possible answer: $(x - 3)(x - 3) = x^2 - 6x + 9$

11. 3

12. $(5x - 3)(5x - 3)$; Find the square root of the first term and the third term. Write the factors in $(a - b)(a - b)$ form.

13. $(6x^2 - 1)(6x^2 - 1)$

Materials List/Setup

Station 1   number cube

Station 2   eight blank index cards; ten index cards with the following written on them:

$3x$; x; $+1$; $+2$; $+4$; $+8$; -1; -2; -4; -8

Station 3   five index cards with the following expressions written on them:

$x^2 + 8x + 12$; $x^2 - 8x + 15$; $x^2 + 2x - 80$; $x^2 + x - 12$; $x^2 - x - 12$

five index cards with the following factors written on them:

$(x - 3)(x + 4)$; $(x + 10)(x - 8)$; $(x + 2)(x + 6)$; $(x + 3)(x - 4)$; $(x - 3)(x - 5)$

Station 4   number cube

Discussion Guide

To support students in reflecting on the activities and to gather some formative information about student learning, use the following prompts to facilitate a class discussion to "debrief" the station activities.

Prompts/Questions

1. How do you find the greatest common factor of terms with variables?

2. How do you factor a trinomial with a leading coefficient not equal to 1?

3. How do you factor a trinomial with a leading coefficient equal to 1?

4. How do you factor the difference of two squares?

5. How do you factor the perfect square trinomial $a^2 + 2ab + b^2$?

6. How do you factor the perfect square trinomial $a^2 - 2ab + b^2$?

Think, Pair, Share

Have students jot down their own responses to questions, then discuss with a partner (who was not in their station group), and then discuss as a whole class.

Suggested Appropriate Responses

1. Find the greatest common factors of the coefficients. Find the variable with the lowest exponent that can be divided into each term of the polynomial.

2. Find the factors of the leading coefficient. Find the factors of the last term that add up to the middle term taking into account the factors of the leading coefficient.

3. Find the factors of the last term that add up to the middle term taking into account x and x as the first terms. (Assuming the first term is x^2.)

4. Take the square root of the first term and the second term. Put in the form $(a - b)(a + b)$.

5. Take the square root of the first term and the third term. Put in the form $(a + b)(a + b)$.

6. Take the square root of the first term and the third term. Put in the form $(a - b)(a - b)$.

Possible Misunderstandings/Mistakes

- Not factoring out the greatest common factor first

- Not using the law of exponents correctly when factoring

- Not finding the factors of the third term that add up to the middle term when factoring trinomials

- Not canceling out the middle term when factoring the difference of two squares

Seeing Structure in Expressions
Set 3: Factoring Polynomials

Station 1

At this station, you will find a number cube. As a group, roll the number cube. Write your answer in the box below. Repeat this process until all the boxes contain a number.

$$x^{\square}, \; x^{\square}, \; x^{\square}$$

1. Of the three terms above, which term has the lowest exponent? _____

2. Divide your answer from problem 1 into each of the three terms above. Write your answers below.

    _____

    _____

    _____

3. You found the largest monomial that could be divided into all the terms. What is the name for this factor?

    _____

Roll the number cube to populate the boxes for each problem below.

4. $4x^{\square} - 6x^{\square} + 4x^{\square}$

    What is the greatest common factor of the coefficients? _____

    What is the greatest common factor of the variables? _____

    What is the greatest common factor of the three terms? _____

    Factor out the greatest common factor of each term. Show your work.

continued

© 2011 Walch Education

Algebra I Station Activities for Common Core State Standards

5. $-5x^{\square}y + x^{\square} - 10x^{\square}y^{\square}$

    What is the greatest common factor of the coefficients? _____

    What is the greatest common factor of the variables? _____

    What is the greatest common factor of the three terms? _____

    Factor out the greatest common factor of each term. Show your work.

6. $6x^{\square}yz + 2s^{\square} + 4x^{\square}y^{\square}$

    What is the greatest common factor of the coefficients? _____

    What is the greatest common factor of the variables? _____

    What is the greatest common factor of the three terms? _____

    Factor out the greatest common factor of each term. Show your work.

7. Did you factor out any variables in problem 6? Why or why not?

8. What is the greatest common factor of the following equation?

    $12x^4y^5z^3 + 56x^2y^4z^6 - 24x^3y^7z^8$

    _____

continued

9. What is the greatest common factor of the following equation?

   $27a^2b^3c^2 - 12ac^3 - 9b^2c^5$

   _____

10. What is the greatest common factor of the following equation?

    $-36r^2s^3t^2 - 18rs^3t - 27r^2s^4t^5 - 9s^2t$

    _____

Seeing Structure in Expressions
Set 3: Factoring Polynomials

Station 2

At this station, you will find eight blank index cards, plus ten index cards with the following written on them:

$3x$; x; $+1$; $+2$; $+4$; $+8$; -1; -2; -4; -8

As a group, determine which index cards to use to factor:

$3x^2 + 14x + 8$

1. What are the factors of $3x^2 + 14x + 8$? _____

2. How did you determine which index cards to use in problem 1?

3. Why were $3x$ and x the only factors of $3x^2$?

Given: $6x^2 + 7x - 5$

4. What are the factors of $6x^2$? _____
   Write each factor on separate index cards.

5. What are the factors of -5? _____
   Write each factor on separate index cards.

continued

Seeing Structure in Expressions
Set 3: Factoring Polynomials

6. As group, arrange the index cards you created to help you factor $6x^2 + 7x - 5$. Show your work.

7. How can you double-check to see if you factored the trinomial correctly?

Station 3

At this station, you will find five index cards with the following expressions written on them:

$$x^2 + 8x + 12;\ x^2 - 8x + 15;\ x^2 + 2x - 80;\ x^2 + x - 12;\ x^2 - x - 12$$

You will also find five index cards with the following factors written on them:

$$(x - 3)(x + 4);\ (x + 10)(x - 8);\ (x + 2)(x + 6);\ (x + 3)(x - 4);\ (x - 3)(x - 5)$$

Shuffle the cards. As a group, match the expressions with their factors. Write the matches on the lines below.

1. _____

2. _____

3. _____

4. _____

5. _____

6. What strategy did you use to match the cards?

7. How can you double-check your matches?

continued

Seeing Structure in Expressions
Set 3: Factoring Polynomials

Given: $x^2 + 4x - 12$

8. What is the greatest common factor of all three terms? _____

Use your answers for problems 9 and 10 to fill in the boxes below.

$(\square + \square)(\square + \square)$

9. What are the factors of x^2? _____

   Write these factors in the solid boxes.

10. What are the factors of -12 that add up to 4? _____

    Write these factors in the dashed boxes.

    The factors of $x^2 + 4x - 12$ are _____.

Given: $2x^2 + 8x - 10$

11. What is the greatest common factor of all three terms? _____

12. Factor out the greatest common factor. What is the new expression?

    _____

Use your answers for problems 13 and 14 to fill in the boxes below.

$(\square + \square)(\square + \square)$

13. What are the factors of x^2? _____

    Write these factors in the solid boxes.

continued

Seeing Structure in Expressions
Set 3: Factoring Polynomials

14. What are the factors of _____ that add up to _____?

    Write these factors in the dashed boxes.

    The three factors of $2x^2 + 8x - 10$ are _____.

15. Why should you factor out the greatest common factor first before factoring the expression?

Seeing Structure in Expressions
Set 3: Factoring Polynomials

Station 4

At this station, you will find a number cube. As a group, roll the number cube once. Write this number in *both* boxes below.

$$(x + \boxed{})(x - \boxed{})$$

1. Use the distribution method to multiply the two binomials. Show your work.

2. How many terms does the polynomial you created in problem 1 contain?

    _____

3. Why is there no x term in the polynomial you created in problem 1?

4. How can you factor $4x^2 - 9$ using the observations you made in problems 1–3? Show your work.

5. Factor $49x^6 - 36$. Show your work.

continued

Seeing Structure in Expressions
Set 3: Factoring Polynomials

As a group, roll the number cube once. Write this number in *both* boxes below.

$$(\boxed{}x + 3)(\boxed{}x + 3)$$

6. Use the distribution method to multiply the two binomials. Show your work.

7. How many terms does the polynomial you created in problem 6 contain?

   _____

8. How can you factor $16x^2 + 24x + 9$ using the observations you made in problems 6 and 7? Show your work.

9. Factor $4x^8 + 20x^4 + 25$. Show your work.

continued

Seeing Structure in Expressions
Set 3: Factoring Polynomials

As a group, roll the number cube once. Write this number in *both* boxes below.

$$(x - \boxed{})(x - \boxed{})$$

10. Use the distribution method to multiply the two binomials. Show your work.

11. How many terms does the polynomial you created in problem 10 contain?

    _____

12. How can you factor $25x^2 - 30x + 9$ using the observations you made in problems 10 and 11? Show your work and answer.

13. Factor $36x^4 - 12x^2 + 1$. Show your work.

Creating Equations

Set 1: Literal Equations

Goal: To provide opportunities for students to develop concepts and skills related to solving literal equations for a specified variable

Common Core Standards

Algebra: Creating Equations★

Create equations that describe numbers or relationships.

A-CED.4. Rearrange formulas to highlight a quantity of interest, using the same reasoning as in solving equations.

Algebra: Reasoning with Equations and Inequalities

Solve equations and inequalities in one variable.

A-REI.3. Solve linear equations and inequalities in one variable, including equations with coefficients represented by letters.

Student Activities Overview and Answer Key

Station 1

Students will be given five index cards with the following formulas and equations written on them:

$$y = mx + b \qquad d = rt \qquad A = lw \qquad V = lwh \qquad A = \frac{1}{2}bh$$

Students work as a group to match the formula or equation with the appropriate "Solve for" variable card and solve for the variable.

Answers

1.

Original problem	Steps	Final answer
$y = mx + b$	$y = mx + b$ $y - b = mx$ $\dfrac{y - b}{m} = x$	$x = \dfrac{y - b}{m}$

(continued)

Original problem	Steps	Final answer
$d = rt$	$d = rt$ $\dfrac{d}{t} = r$	$r = \dfrac{d}{t}$
$A = lw$	$A = lw$ $\dfrac{A}{l} = w$	$w = \dfrac{A}{l}$
$V = lwh$	$V = lwh$ $\dfrac{V}{wh} = l$	$\dfrac{V}{wh} = l$
$A = \dfrac{1}{2}bh$	$A = \dfrac{1}{2}bh$ $\dfrac{2}{1}A = bh$ $\dfrac{2A}{b} = h$	$h = \dfrac{2A}{b}$

2. $x = \dfrac{25 - y}{3}$

 $y = -3x + 25$

3. $a = \dfrac{7 - 28c}{14}$ or $a = \dfrac{1}{2} - 2c$

 $c = \dfrac{7 - 14a}{28}$ or $c = \dfrac{1}{4} - \dfrac{1}{2}a$

4. $g = 12f - \dfrac{3}{2}h + 20$

 $f = \dfrac{1}{12}g + \dfrac{1}{8}h - \dfrac{5}{3}$

 $h = \dfrac{-2}{3}g + 8f + \dfrac{40}{3}$

Station 2

Students are given a number cube. Students roll the number cube to populate the coefficients of variables in equations. Then they solve for each variable in the equation.

Answers

1. Answers will vary. Possible answers include: $6x - 4y = 6z$; $x = \dfrac{2}{3}y + z$; $y = \dfrac{3}{2}x - \dfrac{3}{2}z$; $z = x - \dfrac{2}{3}y$

2. Answers will vary. Possible answers include: $\dfrac{2}{3}A + B = 6C$; $A = \dfrac{-3}{2}B + 9C$; $B = \dfrac{-2}{3}A + 6C$; $C = \dfrac{1}{9}A + \dfrac{1}{6}B$

3. Answers will vary. Possible answers include: $f^2 - 2d = \dfrac{3}{4}g$; $f = \pm\sqrt{2d + \dfrac{3}{4}g}$; $d = \dfrac{1}{2}f^2 - \dfrac{3}{8}g$; $g = \dfrac{4}{3}f^2 - \dfrac{8}{3}d$

Station 3

Students will solve four real-world applications of literal equations. They will be given six formulas to choose from. They will be given a word problem and asked to solve for a specified variable.

Answers

1. $d = rt$; $750 = 60t$; $t = 12.5$ hours; $d = rt$; we solved for t because we wanted to know how long it took him to drive to the college.

2. $V = lwh$; $1280 = 16 \bullet w \bullet 10$; $w = 8$ inches; $V = lwh$; we are given the volume of the aquarium.

3. $A = \dfrac{1}{2}bh$; $6 = \dfrac{1}{2}b(4)$; $b = 3$ inches; $A = \dfrac{1}{2}bh$; we know it is a triangular sail.

4. $A = \pi r^2$; $36\pi = \pi r^2$; $r = 6$ inches; $A = \pi r^2$; we know it is a circular piece of wood.

Creating Equations
Set 1: Literal Equations

Station 4

Students will be given 37 index cards with variables and operations written on them. Students work together to move the cards on the table to create each equation and solve for each variable indicated.

Answers

1. $X = D - Y$

2. $A = TS - B$

3. $R = \dfrac{-A}{F - Z}$

4. $T = X + \dfrac{Z}{S} + \dfrac{A}{S}$ or $T = \dfrac{Z + A}{S} + X$

Materials List/Setup

Station 1   five index cards with the following formulas and equations written on them:

$$y = mx + b,\ d = rt,\ A = lw,\ V = lwh,\ A = \frac{1}{2}bh$$

five index cards with the following written on them:

"Solve for *x*," "Solve for *r*," "Solve for *w*," "Solve for *l*," "Solve for *h*"

Station 2   number cube

Station 3   none

Station 4   37 index cards with the following written on them:

$$X, Y, D, F, A, B, S, T, Z, R, -Y, -Y, (S), (S), -B, -B, -Z, -Z, (R), (R), \left(\frac{1}{F-Z}\right),$$
$$\left(\frac{1}{F-Z}\right), \left(\frac{1}{S}\right), \left(\frac{1}{S}\right), +X, +X, (D), (D), -T, -T, (, (,),), +, -, =$$

Discussion Guide

To support students in reflecting on the activities and to gather some formative information about student learning, use the following prompts to facilitate a class discussion to "debrief" the station activities.

Prompts/Questions

1. How do you solve literal equations for a specific variable?

2. How many variables can you solve for in the formula $V = lwh$?

3. Name three formulas in geometry that are literal equations.

4. Name four real-world applications of literal equations.

Think, Pair, Share

Have students jot down their own responses to questions, then discuss with a partner (who was not in their station group), and then discuss as a whole class.

Suggested Appropriate Responses

1. Isolate the variable and make sure it has a coefficient of 1.

2. four variables which include V, l, w, and h

3. Answers will vary. Possible answers include: $P = 2L + 2W$, $a^2 + b^2 = c^2$, $A = lw$, $P = 4S$

4. Answers will vary. Possible answers include: distance formula, $d = rt$; slope of a hill, $y = mx + b$; interest formula, $I = prt$; force and acceleration formula, $F = ma$

Possible Misunderstandings/Mistakes

- Not solving for the specified variable
- Not following the properties of equality when moving terms from each side of the equal sign
- Not keeping track of the appropriate sign of the term
- Using the wrong formula for real-world applications

Creating Equations
Set 1: Literal Equations

Station 1

You will be given five index cards with the following formulas and equations written on them:

$y = mx + b$, $d = rt$, $A = lw$, $V = lwh$, $A = \dfrac{1}{2}bh$. You will also get five index cards with the following written on them: "Solve for x," "Solve for r," "Solve for w," "Solve for l," "Solve for h."

  Shuffle the cards and deal a card to each student in your group until all the cards have been dealt. Work as a group to match the formula or equation with the appropriate "Solve for" variable card.

1. For all five formulas/equations, write the original problem, your steps to solving for the appropriate variable, and your final answer.

Original problem	Steps	Final answer

Algebra I Station Activities for Common Core State Standards

Creating Equations
Set 1: Literal Equations

Solve for each variable in the equations that follow.

2. $3x + y = 25$

3. $14a + 28c = 7$

4. $\dfrac{1}{2}g - 6f + \dfrac{3}{4}h = 10$

Creating Equations
Set 1: Literal Equations

Station 2

You will be given a number cube. For each problem, roll the number cube and write the result in the box below. Repeat this process until all the boxes contain a number.

1. Solve for each variable. Write your answers in the space below.

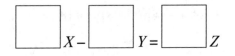

2. Solve for each variable. Write your answers in the space below.

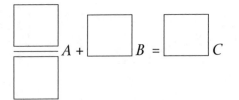

3. Solve for each variable. Write your answers in the space below.

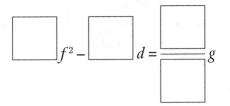

Creating Equations
Set 1: Literal Equations

Station 3

At this station, you will read four real-world applications of literal equations. You will use some of the following formulas to solve the problems:

$$A = \pi r^2 \quad y = mx + b \quad C = 2\pi r \quad d = rt \quad V = lwh \quad A = lw \quad A = \frac{1}{2}bh$$

Work together to read and solve each problem. Write your answers in the space provided.

1. Jesse is going on a road trip to visit a college. The college is 750 miles away.

   How long will it take Jesse to drive to the college if his average speed is 60 miles per hour?

   Which formula did you use to solve this problem and why?

   Which variable did you solve for and why?

2. Amanda has a rectangular fish aquarium that holds 1,280 in³ of water. The length of the aquarium is 16 inches and the height is 10 inches. What is the width of the aquarium?

   Which formula did you use to solve this problem and why?

continued

Creating Equations
Set 1: Literal Equations

3. Matt is building a model sailboat. He wants to construct a triangular sail that has an area of 6 square inches. If the height of the sail is 4 inches, then what is the base of the sail?

   Which formula did you use to solve this problem and why?

4. Sara is going to paint a circular piece of wood for the set of her school play. If the area of the wood is 36π, then what is its radius?

   Which formula did you use to solve this problem and why?

© 2011 Walch Education

Creating Equations
Set 1: Literal Equations

Station 4

You will be given 37 index cards with the following written on them:

$$X, Y, D, F, A, B, S, T, Z, R, -Y, -Y, (S), (S), -B, -B, -Z, -Z, (R), (R), \left(\frac{1}{F-Z} \right),$$
$$\left(\frac{1}{F-Z} \right), \left(\frac{1}{S} \right), \left(\frac{1}{S} \right), +X, +X, (D), (D), -T, -T, (, (,),), +, -, =$$

    Work together to place the cards on the table to create each equation below. Move the cards on the table to solve for each variable indicated. Once you agree on a final answer, write it in the space provided.

1. Solve for X: $X + Y = D$

2. Solve for A: $\dfrac{A + B}{S} = T$

3. Solve for R: $Z - \dfrac{A}{R} = F$

4. Solve for T: $S(T - X) = Z + A$

Creating Equations

Set 2: Graphing Linear Equations/Solving Using Graphs

Goal: To provide opportunities for students to develop concepts and skills related to graphing a line from a table of values, x- and y-intercepts, slope-intercept form, standard form, and point-slope form. Students will also determine the slope and x- and y-intercepts given a graph or two points on the line.

Common Core Standards

Algebra: Creating Equations★

Create equations that describe numbers or relationships.

A-CED.2. Create equations in two or more variables to represent relationships between quantities; graph equations on coordinate axes with labels and scales.

Algebra: Reasoning with Equations and Inequalities

Represent and solve equations and inequalities graphically.

A-REI.10. Understand that the graph of an equation in two variables is the set of all its solutions plotted in the coordinate plane, often forming a curve (which could be a line).

Student Activities Overview and Answer Key

Station 1

Students will be given dry spaghetti noodles, graph paper, and a ruler. They will construct a line using the spaghetti noodles from a table of values. They will switch the x- and y-values in the table and construct the line using spaghetti noodles to identify how this graph relates to the original graph. Then they will construct a line using spaghetti noodles given x- and y-intercepts.

Answers

1. The x- and y-values represent the coordinate pairs used to construct the graph.

Creating Equations
Set 2: Graphing Linear Equations/Solving Using Graphs

2.

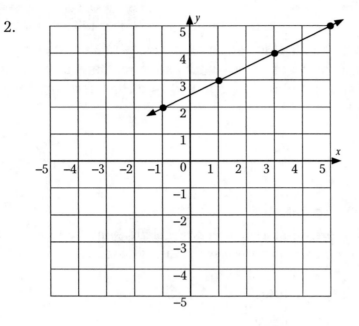

3.

x-value	y-value
2	−1
3	1
4	3
5	5

4.

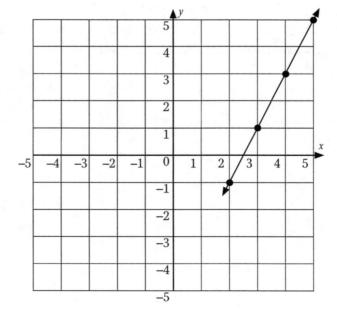

5. No, the graphs do not look the same. The coordinates are different and the slopes are 1/2 versus 2.

6.

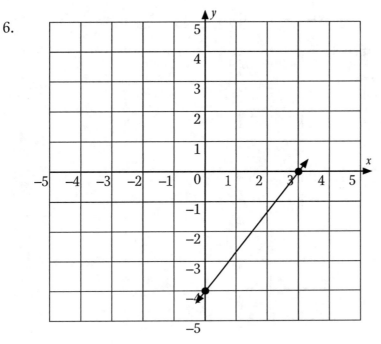

(3, 0) is the x-intercept.

(0, −4) is the y-intercept.

Station 2

Students will be given graph paper and a ruler. They complete a table of x- and y-values given equations in slope-intercept form. Then they graph these equations. They find an equation given the slope and a point on the graph. Then they graph the equation using the x- and y-intercepts.

Answers

1. Answers vary. Sample answers: (0, 3), (1, 5), (2, 7)

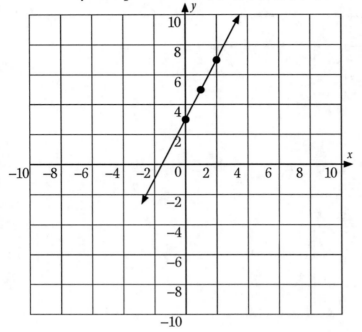

2. Answers vary. Sample answers: (–3, –4) (0, –5), (3, –6)

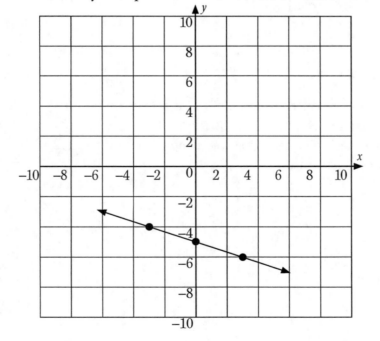

3.

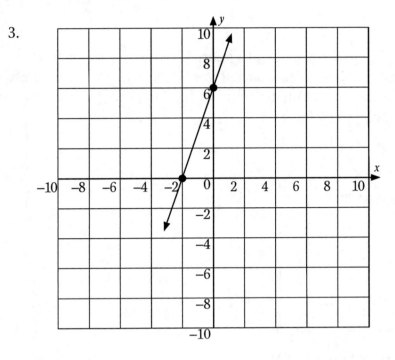

Use the x- and y-intercepts, slope, and/or point $(-1, 3)$ to find the equation.

4. $y = 3x + 6$

5.

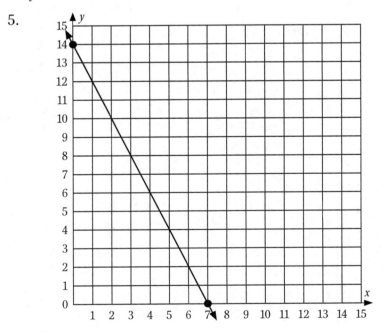

Use the x- and y-intercepts, slope, and/or point $(4, 6)$ to find the equation.

6. $y = -2x + 14$

Station 3

Students will be given spaghetti noodles, graph paper, and a ruler. They will be given a linear equation in standard form and will rewrite it in slope-intercept form. Then they will graph the linear equation. They will be given the slope and a point on the graph and graph it using a spaghetti noodle. Then they will write the equation in point-slope form.

Answers

1. $y = -3x + 2$

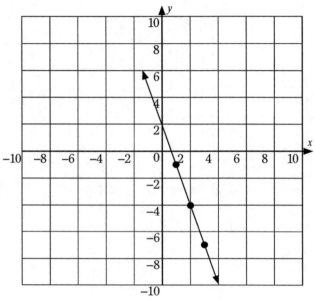

2. $y = \frac{1}{4}x - 3$

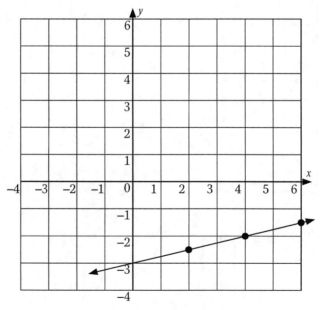

3. $y = 5x - 10$

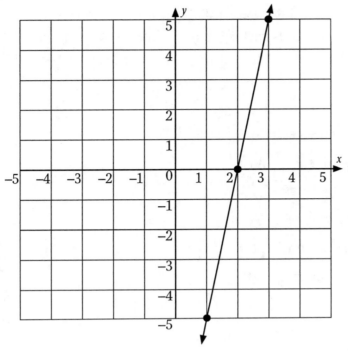

4. $y = -\frac{1}{2}x - \frac{1}{2}$

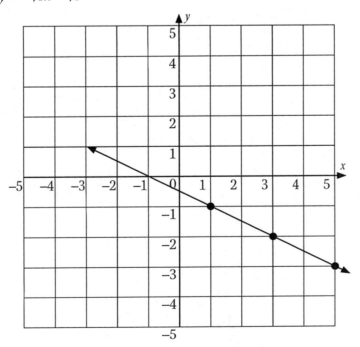

Algebra I Station Activities for Common Core State Standards

Station 4

Students will be given graph paper and a ruler. Students will determine the slope, x- and y-intercepts given a graph of a linear equation. They will construct a graph of a line given two points. Then they will find the slope and x- and y-intercepts of the line.

Answers

1. slope = $1/2$; x-intercept = $(2, 0)$; y-intercept = $(0, -1)$
2. slope = -1; x-intercept = $(3, 0)$, y-intercept = $(0, 3)$
3. slope = 0; x-intercept = none; y-intercept = $(0, 2)$
4. slope = $1/2$; x-intercept = $(0, 0)$; y-intercept = $(0, 0)$
5. slope = $-5/6$; x-intercept = $(8/5, 0)$; y-intercept = $(0, 4/3)$
6. slope = -4; x-intercept = $(5/2, 0)$; y-intercept = $(0, 10)$

Materials List/Setup

Station 1   spaghetti noodles; graph paper; ruler

Station 2   spaghetti noodles; number cube; graph paper; ruler

Station 3   spaghetti noodles; graph paper; ruler

Station 4   graph paper; ruler

Discussion Guide

To support students in reflecting on the activities and to gather some formative information about student learning, use the following prompts to facilitate a class discussion to "debrief" the station activities.

Prompts/Questions

1. How can you find the y-value coordinate given an equation and its x-coordinate?

2. What are the definitions of the terms "x-intercept" and "y-intercept"?

3. How do you write the slope-intercept form of a linear equation?

4. How do you write the standard form of a linear equation?

5. How can you write an equation in slope-intercept form if it is given to you in standard form?

6. How can you find the equation for a line given two points on the line?

Think, Pair, Share

Have students jot down their own responses to questions, then discuss with a partner (who was not in their station group), and then discuss as a whole class.

Suggested Appropriate Responses

1. Plug the x-value into the equation and solve for the y-coordinate.

2. The x-intercept is the coordinate pair at which the graph crosses the x-axis. The y-intercept is the coordinate pair at which the graph crosses the y-axis.

3. $y = mx + b$

4. $ax + by = c$

5. Solve for y and write the equation as $y = mx + b$.

6. Find the slope of the line from the two points. Then plug in the slope and one of the points into $y - y_1 = m(x - x_1)$. Then solve for y.

Possible Misunderstandings/Mistakes

*   Incorrectly plotting the coordinates in reverse order

*   Not keeping track of signs when writing equations in slope-intercept and point-slope forms

*   Not making sure that the y-value is 0 for the x-intercept

*   Not making sure that the x-value is 0 for the y-intercept

*   Incorrectly finding the slope as run/rise instead of rise/run

Creating Equations
Set 2: Graphing Linear Equations/Solving Using Graphs

Station 1

You will be given spaghetti noodles, graph paper, and a ruler. Use these materials and the information in the table below to answer the problems and construct the graphs.

x-value	y-value
−1	2
1	3
3	4
5	5

1. What would the x- and y-values in the table represent if you used them to graph a line?

2. Plot the table of values on your graph paper. Place the spaghetti noodles through all the points to create a line that represents the graph.

3. For each row in the table, switch the x-values with the y-values. Write the new values in the table below.

x-value	y-value

4. On the same graph, plot the table of values on your graph paper. Place the spaghetti noodles through all the points to create a line that represents the graph.

continued

Creating Equations
Set 2: Graphing Linear Equations/Solving Using Graphs

5. Does your graph look the same as problem 2? Why or why not?

6. A graph contains points (3, 0) and (0, –4). Use your graph paper to plot these points and construct a line through the points using the spaghetti noodles.

   These coordinates have special names.

   What is the point (3, 0) called? _____

   What is the point (0, –4) called? _____

NAME:

Creating Equations
Set 2: Graphing Linear Equations/Solving Using Graphs

Station 2

You will be given a number cube, spaghetti noodles, graph paper, and a ruler. For problems 1 and 2, you are given an equation in slope-intercept form.

1. As a group, roll the number cube. Write the result in the first row of the *x*-value column below. Repeat this process until all the rows of the *x*-value contain a number.

   Work together to complete the table of *x*- and *y*-values based on the equation $y = 2x + 3$.

x-value	*y*-value

   As a group, graph the equation on your graph paper.

2. As a group, roll the number cube. Write the result in the first row of the *x*-values column below. Repeat this process until all the rows of the *x*-values contain a number.

   Work together to complete the table of *x*- and *y*-values based on the equation $y = -\dfrac{1}{3}x - 5$.

x-value	*y*-value

   As a group, graph the equation on your graph paper.

3. Use a spaghetti noodle to graph a line that has a slope of 3 and passes through the point (–1, 3). How can you use this graph to find the equation of the line?

continued

Creating Equations
Set 2: Graphing Linear Equations/Solving Using Graphs

4. Write the equation of this line in slope-intercept form. (*Hint*: Use $y = mx + b$.)

   _____

5. Use a spaghetti noodle to graph a line that has a slope of –2 and passes through the point (4, 6). How can you use this graph to find the equation of the line?

6. Write an equation of this line in slope-intercept form. (*Hint*: Use $y = mx + b$.)

   _____

© 2011 Walch Education

Creating Equations
Set 2: Graphing Linear Equations/Solving Using Graphs

Station 3

You will be given spaghetti noodles, graph paper, and a ruler. For problems 1 and 2, you will be given an equation in standard form.

1. Work together to write the following equation in slope-intercept form: $9x + 3y = 6$

   _____

   Determine three coordinate pairs that are on the graph of $9x + 3y = 6$. Write your calculations and coordinate pairs in the space below.

   As a group, graph the equation using graph paper.

2. Work together to write the following equation in slope-intercept form: $2x - 8y = 24$

   _____

   Determine three coordinate pairs that are on the graph of $2x - 8y = 24$. Write your calculations and coordinate pairs in the space below.

   As a group, graph the equation using graph paper.

continued

Creating Equations
Set 2: Graphing Linear Equations/Solving Using Graphs

For problems 3 and 4, you will be given a linear equation written in point-slope form:

$$y - y_1 = m(x - x_1)$$

3. On your graph paper, use a spaghetti noodle to graph the line $y - 10 = 5(x - 4)$.

   Use the graph to find the equation of the line in slope-intercept form.

   Convert the equation from point-slope form into slope-intercept form. Show your work below.

4. On your graph paper, use a spaghetti noodle to graph the line $y + 2 = -\frac{1}{2}(x - 3)$.

   Use the graph to find the equation of the line in slope-intercept form.

   Convert the equation from point-slope form into slope-intercept form. Show your work below.

Creating Equations
Set 2: Graphing Linear Equations/Solving Using Graphs

Station 4

You will be given graph paper, a ruler, and graphs of linear equations. For problems 1–3, work together to determine the slope and *x*- and *y*-intercepts.

1.

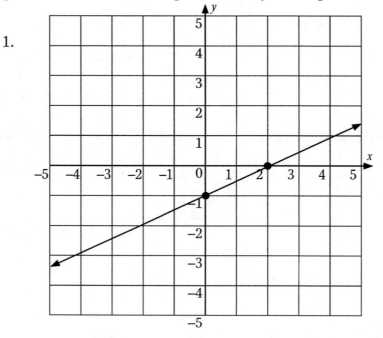

What is the slope of the graph? _____

What is the *x*-intercept? _____

What is the *y*-intercept? _____

2.

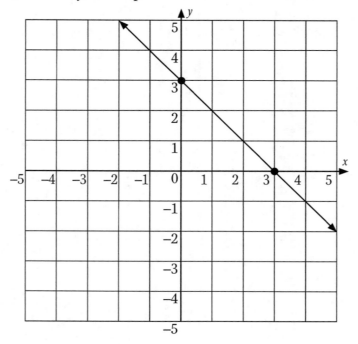

continued

Creating Equations
Set 2: Graphing Linear Equations/Solving Using Graphs

What is the slope of the graph? _____

What is the *x*-intercept? _____

What is the *y*-intercept? _____

3.

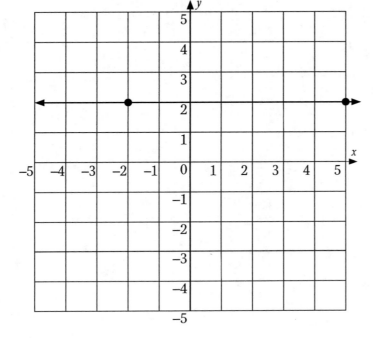

What is the slope of the graph? _____

What is the *x*-intercept? _____

What is the *y*-intercept? _____

For problems 4–6, work together to determine the slope, *x*-intercept, and *y*-intercept when given two points on the line. As a group, plot each set of two points on graph paper and construct a straight line through the points.

4. (8, 4), (12, 6)

What is the slope of the graph? _____

What is the *x*-intercept? _____

What is the *y*-intercept? _____

continued

Creating Equations
Set 2: Graphing Linear Equations/Solving Using Graphs

5. (–2, 3), (4, –2)

   What is the slope of the graph? _____

   What is the x-intercept? _____

   What is the y-intercept? _____

6. (1, 6), (–2, 18)

   What is the slope of the graph? _____

   What is the x-intercept? _____

   What is the y-intercept? _____

Creating Equations

Set 3: Writing Linear Equations

Goal: To provide opportunities for students to develop concepts and skills related to writing linear equations in slope-intercept and standard form given two points and a point and slope

Common Core Standards

Algebra: Creating Equations★

Create equations that describe numbers or relationships.

A-CED.2. Create equations in two or more variables to represent relationships between quantities; graph equations on coordinate axes with labels and scales.

Algebra: Reasoning with Equations and Inequalities

Represent and solve equations and inequalities graphically.

A-REI.10. Understand that the graph of an equation in two variables is the set of all its solutions plotted in the coordinate plane, often forming a curve (which could be a line).

Student Activities Overview and Answer Key

Station 1

Students will be given spaghetti noodles, graph paper, and the general point-slope form of an equation. Students will use the spaghetti noodles to model different equations. Then they will write the slope-intercept form of an equation given the slope and y-intercept. They will also write the slope-intercept form of an equation given the y-intercept and a parallel or perpendicular equation.

Answers

1. $y = 5x + 10$
2. $y = -4x - 8$
3. $y = \dfrac{-1}{3}x + 18$
4. $y = -4x$

Station 2

Students will be given spaghetti noodles, graph paper, and the general slope-intercept and point-slope forms of a linear equation. They will determine which form of the equation to use first in order to write a linear equation when given the slope and a point on the line. They will use spaghetti noodles to model the equations. Then they will write the slope-intercept form of a line given the slope and a point on the line.

Answers

1. Use point-slope form first because you can plug in the given point for x_1 and y_1.
2. $y = 2x + 14$
3. $y = \dfrac{1}{8}x + \dfrac{95}{8}$
4. $y = -3x - 22$
5. $y = \dfrac{-2}{5}x + 14$

Station 3

Students will be given a number cube. They will use it to populate two ordered pairs. From the ordered pairs, they will find the slope and write the equation in slope-intercept form. Then they will explain how they found the slope and the equation in slope-intercept form.

Answers

1. Answers will vary.

2. Answers will vary.

3. Answers will vary.

4. Slope $= \dfrac{\text{rise}}{\text{run}} = \dfrac{y_2 - y_1}{x_2 - x_1}$

5. Find the slope from the two points. Plug one of the points and the slope into $y = mx + b$ to find b. Then write the equation in slope-intercept form.

Station 4

Students will be given 10 index cards with the following equations written on them:

$$y = 2x + 12 \qquad y = -5x - 8 \qquad y = \frac{1}{2}x + 16 \qquad y = \frac{-2}{3}x - 20 \qquad y = 10x - 24$$

$$-x + 2y = 32 \qquad 2x + 3y = -60 \qquad 4x - 2y = -24 \qquad -15x + \frac{3}{2}y = -36 \qquad \frac{-5}{2}x - \frac{1}{2}y = 4$$

They will work as a group to match the equations written in standard form with the equations written in slope-intercept form. Then they convert an equation written in standard form to slope-intercept form and vice versa.

Answers

1. $y = \frac{1}{2}x + 16$ and $-x + 2y = 32$

2. $y = \frac{-2}{3}x - 20$ and $2x + 3y = -60$

3. $y = 10x - 24$ and $-15x + \frac{3}{2}y = -36$

4. $4x - 2y = -24$ and $y = 2x + 12$

5. $\frac{-5}{2}x - \frac{1}{2}y = 4$ and $y = -5x - 8$

6. $y = \frac{1}{3}x + \frac{4}{3}$

7. $x - 5y = 50$

Materials List/Setup

Station 1   spaghetti noodles; graph paper

Station 2   spaghetti noodles; graph paper

Station 3   number cube

Station 4   10 index cards with the following equations written on them:

$$y = 2x + 12 \qquad y = -5x - 8 \qquad y = \frac{1}{2}x + 16 \qquad y = \frac{-2}{3}x - 20 \qquad y = 10x - 24$$

$$-x + 2y = 32 \qquad 2x + 3y = -60 \qquad 4x - 2y = -24 \qquad -15x + \frac{3}{2}y = -36 \qquad \frac{-5}{2}x - \frac{1}{2}y = 4$$

Discussion Guide

To support students in reflecting on the activities and to gather some formative information about student learning, use the following prompts to facilitate a class discussion to "debrief" the station activities.

Prompts/Questions

1. How do you write the standard form of a linear equation?

2. How do you write the slope-intercept form of a linear equation?

3. How do you write the slope-intercept form of a linear equation given a point and the slope?

4. How do you write the slope-intercept form of a linear equation given two points on the line?

5. How do you convert the standard form of a linear equation into slope-intercept form?

6. How do you convert the slope-intercept form of a linear equation into standard form?

Think, Pair, Share

Have students jot down their own responses to questions, then discuss with a partner (who was not in their station group), and then discuss as a whole class.

Suggested Appropriate Responses

1. $ax + by = c$

2. $y = mx + b$

3. Use the point-slope form of the equation to plug in the given point and slope. Then write the equation in slope-intercept form.

4. Find the slope from the two points. Plug one of the points and the slope into $y = mx + b$ to find b. Then write the equation in slope-intercept form.

5. Solve for y.

6. Move the x-value to the left-hand side of the equation by using either the addition or subtraction property of equality and make sure that the coefficients are whole numbers with the coefficient of x being positive.

Possible Misunderstandings/Mistakes

- Mixing up the x- and y-values when plugging them into the point-slope or slope-intercept forms

- Not keeping track of positive and negative signs when converting from standard form to slope-intercept form and vice versa

- Incorrectly finding the slope as $\dfrac{\text{run}}{\text{rise}}$ instead of $\dfrac{\text{rise}}{\text{run}}$

Creating Equations
Set 3: Writing Linear Equations

Station 1

You will find spaghetti noodles and graph paper at this station. Draw a coordinate plane on your graph paper by creating and labeling the x-axis and the y-axis. For each problem, use the given information and the spaghetti to find the slope-intercept form of the equation.

- Slope-intercept form is written as $y = mx + b$.

1. The equation has a y-intercept of $(0, 10)$ and a slope of 5.

   Model this equation using the spaghetti. Then write the equation in slope-intercept form.

2. The equation has a y-intercept of $(0, -8)$ and is parallel to the line $y = -4x - 16$.

   Model this equation using the spaghetti. Then write the equation in slope-intercept form.

3. The equation has a y-intercept of 18 and is perpendicular to the line $y = 3x + 9$.

   Model this equation using the spaghetti. Then write the equation in slope-intercept form.

4. The graph of the equation passes through the origin and is perpendicular to $y = \dfrac{1}{4} x$.

   Model this equation using the spaghetti. Then write the equation in slope-intercept form.

© 2011 Walch Education

Creating Equations
Set 3: Writing Linear Equations

Station 2

You will find spaghetti noodles and graph paper at this station. Draw a coordinate plane on your graph paper by creating and labeling the x-axis and the y-axis. Use the spaghetti and what you know about the slope-intercept and point-slope forms of a linear equation to answer the questions.

- Slope-intercept form is written as $y = mx + b$.
- Point-slope form is written as $y - y_1 = m(x - x_1)$.

1. You want to find a linear equation. Which form should you use first if you are given the slope and a point on the line?

2. Use spaghetti to model a line that has a slope of 2 and passes through (–3, 8). Then write the equation for this line in slope-intercept form.

3. Use spaghetti to model a line that has a slope of $\frac{1}{8}$ and passes through (1, 12). Then write the equation for this line in slope-intercept form.

4. Use spaghetti to model a line that has a slope of –3 and passes through (–4, –10). Then write the equation for this line in slope-intercept form.

5. Use spaghetti to model a line that has a slope of $-\frac{2}{5}$ and passes through (–15, 20). Then write the equation for this line in slope-intercept form.

Algebra I Station Activities for Common Core State Standards

Creating Equations
Set 3: Writing Linear Equations

Station 3

You will be given a number cube. For each problem, roll the number cube four times to create two ordered pairs. Then use these ordered pairs to find the slope and write the equation in slope-intercept form.

- Slope-intercept form is written as $y = mx + b$.

1. First ordered pair: (,)

   Second ordered pair: (,)

   Slope = _____

   Equation written in slope-intercept form: _____

2. First ordered pair: (,)

   Second ordered pair: (,)

   Slope = _____

   Equation written in slope-intercept form: _____

3. First ordered pair: (,)

   Second ordered pair: (,)

   Slope = _____

   Equation written in slope-intercept form: _____

4. How did you find the slope from the two ordered pairs?

5. How did you find b in the slope-intercept form of the equation?

Creating Equations
Set 3: Writing Linear Equations

Station 4

At this station, you will find 10 index cards with the following equations written on them:

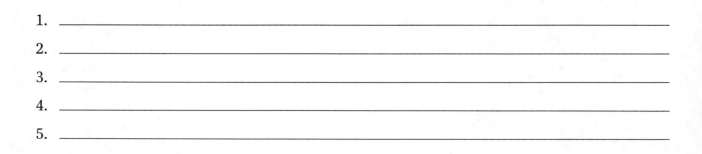

$$y = 2x + 12 \qquad y = -5x - 8 \qquad y = \frac{1}{2}x + 16 \qquad y = \frac{-2}{3}x - 20 \qquad y = 10x - 24$$

$$-x + 2y = 32 \qquad 2x + 3y = -60 \qquad 4x - 2y = -24 \qquad -15x + \frac{3}{2}y = -36 \qquad \frac{-5}{2}x - \frac{1}{2}y = 4$$

As a group, shuffle the cards. Work as a group to match the linear equation written in standard form with the same equation written in slope-intercept form. Write your matches on the lines below.

1. _____

2. _____

3. _____

4. _____

5. _____

6. Rewrite the equation $-3x + 9y = 12$ so that it's in slope-intercept form. Show your work.

7. Rewrite the equation $y = \frac{1}{5}x - 10$ so that it's in standard form. Show your work.

Reasoning with Equations and Inequalities

Set 1: Solving Linear Equations

Goal: To provide opportunities for students to develop concepts and skills related to simplifying algebraic expressions, solving linear equations in one variable, and solving multi-step linear equations and inequalities. Students will also identify and apply properties of real numbers and equality.

Common Core Standards

Algebra: Creating Equations★

Create equations that describe numbers or relationships.

A-CED.1.   Create equations and inequalities in one variable and use them to solve problems.

Algebra: Reasoning with Equations and Inequalities

Understand solving equations as a process of reasoning and explain the reasoning.

A-REI.1.   Explain each step in solving a simple equation as following from the equality of numbers asserted at the previous step, starting from the assumption that the original equation has a solution. Construct a viable argument to justify a solution method.

Solve equations and inequalities in one variable.

A-REI.3.   Solve linear equations and inequalities in one variable, including equations with coefficients represented by letters.

Student Activities Overview and Answer Key

Station 1

Students will be given four index cards that contain equations and fifteen index cards that contain the "steps" required to solve the equations. Students work as a group to place "step" cards sequentially with the appropriate equation, ending in the solution.

Answers

Order of equation cards will vary. Check to ensure steps are grouped with their respective equations in the order found on the following page:

Equation card: $3x + 2 - x = -\dfrac{2}{3}x + 26$

Step cards:

$2x + 2 = -\dfrac{2}{3}x + 26$

$\dfrac{8}{3}x + 2 = 26$

$\dfrac{8}{3}x = 24$

$x = 9$

Equation: $\dfrac{x}{2} = \dfrac{x}{4} + 12$

Step cards:

$\dfrac{x}{2} - \dfrac{x}{4} = 12$

$\dfrac{2x}{4} - \dfrac{x}{4} = 12$

$\dfrac{x}{4} = 12$

$x = 48$

Equation: $10x - 5 = 5x + 20$

Step cards:

$5x - 5 = 20$

$5x = 25$

$x = 5$

Equation: $2(x + 4) = \dfrac{2}{3}(x - 24)$

Step cards:

$2x + 8 = \dfrac{2}{3}x - 16$

$\dfrac{4}{3}x + 8 = -16$

$\dfrac{4}{3}x = -24$

$x = -18$

Station 2

Students will be given eight index cards with the following final simplified algebraic expressions written on them: $4x + 2$, $3b - 10$, $\dfrac{5}{8}y - 12$, $\dfrac{-3}{7}c + \dfrac{4}{9}$, $-x$, $6xy^3$, $\dfrac{2x}{3z^2}$, a^3b^2. Students shuffle and place the index cards in a pile. One student draws a card and they work as a group to create an algebraic expression that precedes this simplified expression. Then they simplify given algebraic expressions.

Answers

1. Answers will vary. Possible answers include:

   $2x + 2x + 2$; $10b - 7b - 5 - 5$; $\dfrac{3}{8}y + \dfrac{1}{4}y - 12$; $\dfrac{-3}{7}c + \dfrac{2}{9} + \dfrac{2}{9}$; $5x - 6x$; $2xy \bullet 3y^2$;

   $\dfrac{x}{3z} \bullet \dfrac{2}{z}$; $a^2b \bullet ab$

2a. $-4x + 15$

2b. $8x^2y + 72y - 3$

2c. $14x^3 + 3xy^2 + 12x^2 - 21x$

2d. $-\dfrac{1}{4}x + 4$

Station 3

Students will be given a number cube. They roll the number cube to populate boxes that will represent the commutative and associative properties. They will derive the commutative and associative properties based on their observations. Students will realize that these properties only work for addition and multiplication. They will derive the distributive property based on an example.

Answers

1. Answers will vary; yes; commutative property is when you change the order of the numbers without changing the result; just addition and multiplication. Subtraction and division won't give you the same answer.

2. Answers will vary; no, because they are addition and multiplication; associative property is when you change the grouping of the numbers without changing the result; just addition and multiplication. Subtraction and division won't give you the same answer.

3. Distributive property is when a number is multiplied by the sum of two other numbers; the first number can be distributed to both of those two numbers and multiplied by each of them separately.

4. a. $10x + 2y$; distributive; b. commutative; c. no, because it is subtraction; d. associative; e. no, because it is division.

5. Answers will vary. Possible answers include: Figuring out the cost of a movie based on the number of people in two families that are going to the movie. $C(F_1 + F_2)$, for which C = cost of movie, F_1 = number of people in family 1, and F_2 = number of people in family 2.

Station 4

Students will work together to write and solve linear equations based on real-world examples. They will provide the equation and the solution.

Answers

1. $45 + 0.25x = 145$; $x = 400$ minutes

2. $125 + 4x = 157$; $x = \$8$ per can

3. $20 + 10x \le 100$; $x \le 8$; no more than 8 aerobics classes per month

4. $x(16 - 2) \ge 210$; $x \ge 15$; no fewer than 15 DVDs per month

Materials List/Setup

Station 1   four index cards with the following equations written on them:

$$3x + 2 - x = -\frac{2}{3}x + 26, \ 10x - 5 = 5x + 20, \ \frac{x}{2} = \frac{x}{4} + 12, \ 2(x + 4) = \frac{2}{3}(x - 24)$$

15 index cards with the following "steps" written on them:

$$2x + 2 = -\frac{2}{3}x + 26 \qquad\qquad \frac{x}{2} - \frac{x}{4} = 12 \qquad 2x + 8 = \frac{2}{3}x - 16$$

$$\frac{8}{3}x + 2 = 26 \qquad 5x - 5 = 20 \qquad \frac{2x}{4} - \frac{x}{4} = 12 \qquad \frac{4}{3}x + 8 = -16$$

$$\frac{8}{3}x = 24 \qquad\quad 5x = 25 \qquad\qquad \frac{x}{4} = 12 \qquad\qquad \frac{4}{3}x = -24$$

$$x = 9 \qquad\qquad\qquad x = 5 \qquad\qquad\quad x = 48 \qquad\qquad\quad x = -18$$

Station 2   eight index cards with the following final simplified algebraic expressions written on them:

$$4x + 2, \ 3b - 10, \ \frac{5}{8}y - 12, \ \frac{-3}{7}c + \frac{4}{9}, \ -x, \ 6xy^3, \ \frac{2x}{3z^2}, \ a^3b^2$$

Station 3   number cube

Station 4   none

Discussion Guide

To support students in reflecting on the activities and to gather some formative information about student learning, use the following prompts to facilitate a class discussion to "debrief" the station activities.

Prompts/Questions

1. How do you solve linear equations?
2. How do you simplify algebraic expressions?
3. What is the commutative property?
4. What is the associative property?
5. What is the distributive property?
6. What are examples of real-world applications of linear equations?

Think, Pair, Share

Have students jot down their own responses to questions, then discuss with a partner (who was not in their station group), and then discuss as a whole class.

Suggested Appropriate Responses

1. Combine like terms if necessary. Use the properties of equality to get the variable by itself on one side of the equation. Solve for the variable by making its coefficient equal to 1.
2. Combine like terms and use the order of operations.
3. The commutative property is when you change the order of the numbers without changing the result.
4. The associative property is when you change the grouping of the numbers without changing the result.
5. The distributive property is when a number is multiplied by the sum of two other numbers; the first number can be distributed to both of those two numbers and multiplied by each of them separately.
6. Answers will vary. Possible answers include: the cost of services (including water, telephone, and electricity) based on a flat rate plus a usage fee

Possible Misunderstandings/Mistakes

- Incorrectly identifying subtraction and division problems as depicting commutative or associative properties

- Incorrectly identifying the constant versus the coefficient of the variable when writing and solving linear equations and inequalities

- Using the wrong sign (i.e., < instead of >) when writing and solving inequalities

- Simplifying algebraic expressions by combining unlike terms

Reasoning with Equations and Inequalities
Set 1: Solving Linear Equations

Station 1

You will be given four index cards with equations written on them. You will also be given fifteen index cards with steps for solving the equations written on them. Place the "equation" cards in one pile and lay out the "step" cards on your table or desk. Draw a card from the equation deck. Find all the step cards that show the steps for solving that equation. Place them in order underneath the equation, ending with the solution as the last step. Repeat the process for each equation card.

On the lines below, write the equation, the steps to solve it in order, and the corresponding last step (the solution).

Equation 1: _____

_____

_____

_____

_____

Equation 2: _____

_____

_____

_____

_____

Equation 3: _____

_____

_____

_____

_____

Equation 4: _____

_____

_____

_____

_____

Reasoning with Equations and Inequalities
Set 1: Solving Linear Equations

Station 2

You will be given eight index cards with the following final simplified algebraic expressions written on them: $4x + 2$, $3b - 10$, $\dfrac{5}{8}y - 12$, $\dfrac{-3}{7}c + \dfrac{4}{9}$, $-x$, $6xy^3$, $\dfrac{2x}{3z^2}$, a^3b^2

Shuffle and place the index cards in a pile. Have one student draw a card. As a group, create an algebraic expression that precedes this simplified expression. For example, if you had drawn a card that read $3x - 15$, then your answer could be $7x - 4x - 15$ or $8x - 10 - 5x - 5$, etc.

1. As a group, come to an agreement on your answers. Write the final algebraic expression and your answers in the space below.

2. As a group, simplify the following algebraic expressions:

   a. $-x + 14 - 3x + 1 =$ _____

   b. $8y(x^2 + 9) - 3 =$ _____

   c. $12x^2 - 19x + 4xy^2 + 14x^3 - 2x - xy^2 =$ _____

   d. $\dfrac{1}{2}x + 4 - \dfrac{3}{4}x =$ _____

Reasoning with Equations and Inequalities
Set 1: Solving Linear Equations

Station 3

Use the number cube provided to complete and solve the problems below.

1. As a group, roll the number cube and write the result in the first box. Roll again, then write the second number in the second box for each problem. Then find the sum and product.

    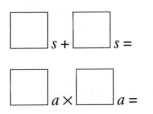

    If you add or multiply the terms in reverse order, do you get the same answers? _____

    This problem represents the commutative property. Write a definition of the commutative property based on your observations.

    Does the commutative property hold for addition, subtraction, multiplication, and division? Why or why not?

2. As a group, roll the number cube and write the result in the first box. Repeat this process to write a number in the second and third boxes for each problem. Then write the same three numbers in the same order in the last three boxes for each problem.

    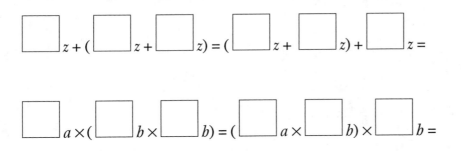

continued

Reasoning with Equations and Inequalities
Set 1: Solving Linear Equations

Does the grouping of the terms, as shown by the parentheses, change the answer on each side of the equation? Why or why not?

This problem represents the associative property. Write a definition of the associative property based on your observations.

Does the associative property hold for addition, subtraction, multiplication, and division? Why or why not?

3. The following problem depicts the distributive property:

$$4x(7 + 9) = 4x(7) + 4x(9) = 28x + 36x = 64x$$

Based on this problem, write a definition for the distributive property in the lines below.

4. For each of the following problems, identify the property used as commutative, associative, or distributive. Simplify the problem if necessary. If the problem doesn't represent one of these properties, explain why.

   a. $2(5x + y) =$

   b. $34 + 12x + 10 = 10 + 34 + 12x$

continued

Reasoning with Equations and Inequalities
Set 1: Solving Linear Equations

c. $12x^2 - 5x^2 = 5x^2 - 12x^2$

d. $(1a + 16a) + 29a = 1a + (16a + 29a)$

e. $24x^2 \div 3y = 3y \div 24x^2$

5. What is one real-world example of when you would use the distributive property?

Reasoning with Equations and Inequalities
Set 1: Solving Linear Equations

Station 4

Write equations for real-world situations and then solve the equations. Work with your group to write and solve the equations. When everyone agrees on the correct equation and solution, write them in the space provided.

1. Janice needs to figure out her cell phone bill. She is charged a monthly flat fee of $45. She is also charged $0.25 per minute. How many minutes did she use if her cell phone bill was $145?

    Equation: _____   Solution: _____

2. Manny spent $157 at a sporting goods store. He bought a warm-up suit for $125 and spent the rest of the money on cans of tennis balls. If each can of tennis balls costs $4, how many cans did he buy?

    Equation: _____   Solution: _____

Now you will write <u>inequalities</u> for real-world situations and then solve the inequalities. Work with your group to write and solve the inequalities.

3. Megan joined a gym for a monthly fee of $20. She has a budget of no more than $100 per month. She wants to take aerobics classes that cost $10. How many aerobics classes can she take each month and stay within her budget?

    Inequality: _____   Solution: _____

4. John sells DVDs on the Internet. He wants to make no less than $210 per month. He sells the DVDs for $16, and it costs him $2 to ship each DVD. How many DVDs must he sell to make no less than $210 per month?

    Inequality: _____   Solution: _____

Reasoning with Equations and Inequalities

Set 2: Real-World Situation Graphs

Goal: To provide opportunities for students to develop concepts and skills related to creating and interpreting graphs representing real-world situations

Common Core Standards

Algebra: Creating Equations★

Create equations that describe numbers or relationships.

A-CED.2. Create equations in two or more variables to represent relationships between quantities; graph equations on coordinate axes with labels and scales.

Algebra: Reasoning with Equations and Inequalities

Represent and solve equations and inequalities graphically.

A-REI.10. Understand that the graph of an equation in two variables is the set of all its solutions plotted in the coordinate plane, often forming a curve (which could be a line).

Student Activities Overview and Answer Key

Station 1

Students will be given a ruler and graph paper. They work together to graph the linear equation of two cell phone company plans. Then they use the graph to compare the two cell phone plans.

Answers

1. $y = 40 + 0.5x$; answers will vary, possible values include:

Minutes (x)	5	10	20	35	45
Cost in $ (y)	42.5	45	50	57.50	62.50

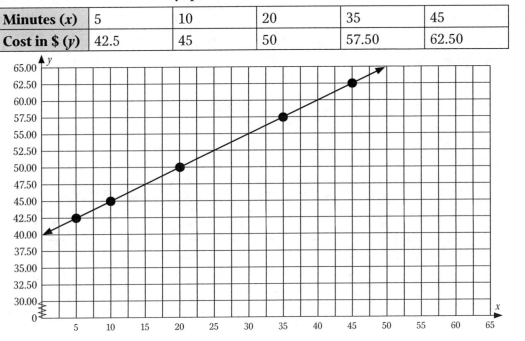

2. $y = 60 + 0.1x$; answers will vary, possible values include:

Minutes (x)	5	10	20	35	45
Cost in $ (y)	60.50	61	62	63.50	64.50

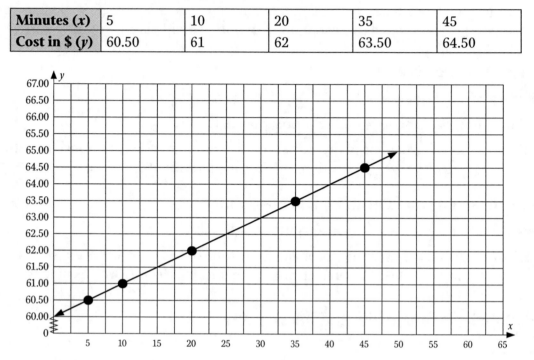

3. They should choose Bell Phone's plan because it only costs $55 versus $63.

4. They should choose Ring Phone's plan because it only costs $68 versus $80.

5. At 50 minutes, it doesn't matter which plan the customer chose because both plans cost $65.

Station 2

Students will be given a ruler and graph paper. They will work together to complete a table of values given an equation, and then graph the equation. They will analyze the slope of the graph as it applies to a real-world value.

Answers

1. $y = 20x + 10$

2. Answers will vary. Possible table of values:

Number of months	Account balance ($)
0	10
2	50
4	90
5	110
8	170
9	190

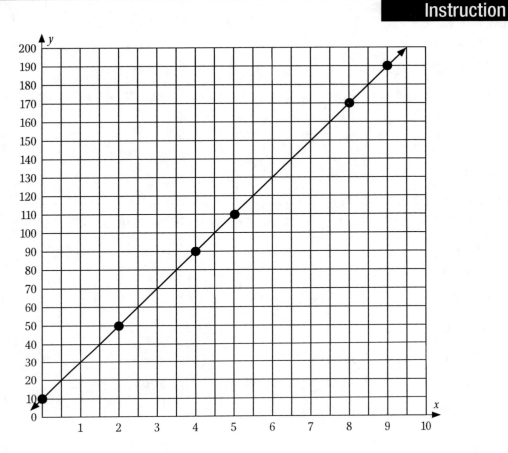

3. He will have $310 in his savings account after 15 months. This will allow him to buy the $300 stereo.

4. The slope of the graph would be steeper because the amount he saves each month represents the slope.

5. The slope of the graph would not be as steep because the amount he saves each month represents the slope.

Station 3

Students will be given a real-world graph of calories burned per mile for runners. They will interpret the graph and explain how to find an equation from the graph.

Answers

1. 60 calories per mile

2. about 69 calories per mile

3. about 81 calories per mile

4. 125 pounds

5. 150 pounds

6. Use two points to find the slope. Use a point and point-slope form to find the equation of the graph.

Station 4

Students will be given a graph that represents the temperature change in the United States in January from 1999–2009. They will analyze the temperature increase and decrease and how it relates to slope.

Answers

1. 2005–06

2. It had the steepest positive slope.

3. 2006–07

4. It had the steepest negative slope.

5. 1999–2000, 2000–01, 2002–03, 2003–04, 2006–07, 2007–08

6. 2001–02, 2004–05, 2005–06, 2008–09

Materials List/Setup

Station 1   graph paper; ruler

Station 2   graph paper; ruler

Station 3   none

Station 4   calculator

Discussion Guide

To support students in reflecting on the activities and to gather some formative information about student learning, use the following prompts to facilitate a class discussion to "debrief" the station activities.

Prompts/Questions

1. Using a graph, how can you find the *x*-value given its *y*-value?

2. Using a graph, how can you find the *y*-value given its *x*-value?

3. Using a graph, how can you find the *x*- and *y*-intercepts of the graph?

4. How can you use an equation to plot its graph?

5. What are examples of real-world situations in which you could construct a graph to represent the data?

6. Do graphs of most real-world situations represent a linear equation? Why or why not?

Think, Pair, Share

Have students jot down their own responses to questions, then discuss with a partner (who was not in their station group), and then discuss as a whole class.

Suggested Appropriate Responses

1. On the graph, move your finger across from the *y*-axis to the line. Move your finger down to the *x*-axis to find the *x*-value.

2. On the graph, move your finger from the *x*-axis up to the line. Move your finger straight across to the *y*-axis to find the *y*-value.

3. The *x*-intercept is where the graph crosses the *x*-axis. The *y*-intercept is where the graph crosses the *y*-axis.

4. Create a table of values that are solutions to the equation. Graph these ordered pairs and draw a line through these points.

5. Answers will vary. Possible answers: Business: yearly revenues; Biology: growth rate; Finance: savings account balance

6. No. Linear equations have a consistent slope. In the real world, the rate of increase or decrease is often variable because of many outside factors.

Possible Misunderstandings/Mistakes

- Reversing the x-values and the y-values when reading the graph
- Incorrectly reading the graph by matching up the wrong x- and y-values
- Reversing the x-values and y-values when constructing the graph
- Incorrectly plugging x-values into the given equation to find the y-values

Reasoning with Equations and Inequalities
Set 2: Real-World Situation Graphs

Station 1

You will be given a ruler and graph paper. Work together to analyze data from the real-world situation described below, then, as a group, answer the questions.

> You are going to get a new cell phone and need to choose between two cell phone companies. Bell Phone Company charges $40 per month. It costs $0.50 per minute after you have gone over the monthly number of minutes included in the plan. Ring Phone Company charges $60 per month. It costs $0.10 per minute after you go over the monthly number of minutes included in the plan.

Let x = minutes used that exceeded the plan. Let y = cost of the plan.

1. Write an equation that represents the cost of the Bell Phone Company's plan.

    _____

    Complete the table by selecting values for x and calculating y.

Minutes (x)					
Cost in $ (y)					

Use your graph paper to graph the ordered pairs. Use your ruler to draw a straight line through the points and complete the graph.

2. Write an equation that represents the cost of Ring Phone Company's plan.

    _____

    Complete the table by selecting values for x and calculating y.

Minutes (x)					
Cost in $ (y)					

continued

Reasoning with Equations and Inequalities
Set 2: Real-World Situation Graphs

On the same graph, plot the ordered pairs. Use your ruler to draw a straight line through the points and complete the graph. Use your graphs to answer the following questions.

3. Which plan should a customer choose if he or she uses 30 minutes of extra time each month? Explain.

4. Which plan should a customer choose if he or she uses 80 minutes of extra time each month? Explain.

5. At what number of extra minutes per month would it not matter which phone plan the customer chose since the cost would be the same? Explain.

Reasoning with Equations and Inequalities
Set 2: Real-World Situation Graphs

Station 2

You will be given a ruler and graph paper. Use the information from the problem scenario below to answer the questions. Let x = months and y = savings account balance.

Marcus is going to start saving $20 every month to buy a stereo. His parents gave him $10 for his birthday to open his savings account.

1. Write an equation that represents Marcus's savings account balance x months after he began saving.

2. Complete the table by selecting variables for x and calculating y.

Number of months	Account balance ($)

Use your graph paper to define the scale of the x- and y-axis and graph the ordered pairs. Use your ruler to draw a straight line through the points and complete the graph.

3. Use your graph to estimate the number of months it will take Marcus to save enough money for a $300 stereo. Explain.

continued

Reasoning with Equations and Inequalities
Set 2: Real-World Situation Graphs

4. If Marcus saved $40 per month instead of $20, how would you expect the slope of the graph to change? Explain.

5. If Marcus saved $10 per month instead of $20, how would you expect the slope of the graph to change? Explain.

Reasoning with Equations and Inequalities
Set 2: Real-World Situation Graphs

Station 3

The equation $y = 0.6x$ represents the number of calories (y) that a runner burns per mile based on their body weight of x pounds.

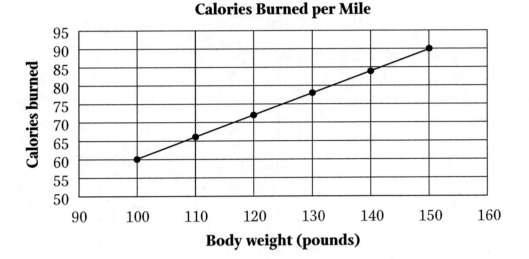

Calories Burned per Mile

For each weight below, use the graph to find the number of calories burned per mile.

1. 100 pounds: _____

2. 115 pounds: _____

3. 135 pounds: _____

For each amount of calories burned per mile below, use the graph to find the matching weight of the person.

4. 75 calories burned: _____

5. 90 calories burned: _____

6. If you didn't know the equation of this graph, how could you use the graph to find the equation of the line? Explain.

Reasoning with Equations and Inequalities
Set 2: Real-World Situation Graphs

Station 4

NOAA Satellite and Information Service created the graph below, which depicts the U.S. National Summary of the temperature in January from 1999–2009.

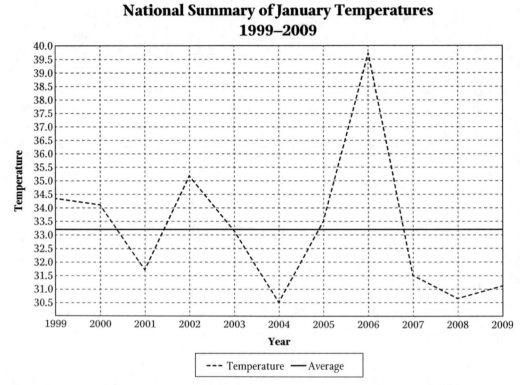

**National Summary of January Temperatures
1999–2009**

Source: www7.ncdc.noaa.gov/CDO/CDODivisionalSelect.jsp#

1. Between which consecutive years did the United States see the greatest increase in average temperature change in January?

2. What strategy did you use to answer problem 1?

continued

Reasoning with Equations and Inequalities
Set 2: Real-World Situation Graphs

3. Between which consecutive years did the United States see the greatest decrease in average temperature change in January?

4. What strategy did you use to answer problem 3?

5. Between which consecutive years was the temperature change represented as a negative slope? Explain.

6. Between which consecutive years was the temperature change represented as a positive slope? Explain.

Reasoning with Equations and Inequalities

Set 3: One-Variable Inequalities

Goal: To provide opportunities for students to develop concepts and skills related to solving single-step, multiple-step, and compound inequalities. They will also graph the inequalities.

Common Core Standards

Algebra: Reasoning with Equations and Inequalities

Solve equations and inequalities in one variable.

A-REI.3. Solve linear equations and inequalities in one variable, including equations with coefficients represented by letters.

Student Activities Overview and Answer Key

Station 1

Students will be given algebra tiles. They will also be given slips of paper with the symbols <, >, ≤, and ≥ written on them. Students work together to model one-step inequalities using the algebra tiles. Then they solve the inequalities. They solve one-step inequalities involving negative numbers. They derive the rule for dealing with negative numbers when solving an inequality. They also graph each inequality.

Answers

1. $x < 5$; yes, because $3 < 5$; no, because 15 is not less than 10; open circle at $x = 5$ because there is not an equal sign in the inequality; number line:

    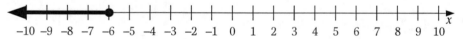

2. No, the solution should be $x \leq -6$ because you divided by a negative number. Also, when $x = 5$, the inequality is $-15 \geq 18$, which is a false statement; closed circle because there is an equal sign in the inequality; number line:

3. No, because you multiplied by a negative number. Also, when $x = -12$, the inequality is $3 < 2$, which is a false statement; open circle, because there is not an equal sign in the inequality; number line:

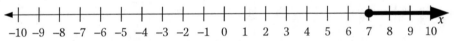

4. When multiplying or dividing by a negative number in an inequality, reverse the original inequality sign to its opposite in order to make the new inequality true.

Station 2

Students will be given multi-step inequalities. They will work as a group to solve the inequalities and graph them on the number line. They will show how multiplying or dividing by a negative number reverses the sign of the inequality.

Answers

1. Add 8 to both sides of the inequality; divide each side by 4; $x \geq 7$; substitute x-values greater than 7 to make sure it is a true statement; closed circle because there is an equal sign in the inequality; number line:

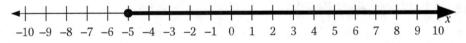

2. Add 19 to both sides of the inequality; divide each side by -10; reverse the sign of the inequality to its opposite; $x \geq -5$; substitute values greater than -5 to make sure it is a true statement; closed circle, because there is an equal sign in the inequality; number line:

3. Subtract $\dfrac{2}{5}x$ from both sides of the inequality; multiply each side by $\dfrac{-5}{3}$; reverse the sign of the inequality to its opposite; $x < -25$; substitute x-values less than -25 to make sure it is a true statement; open circle, because there is not an equal sign in the inequality; number line:

133

Station 3

Students will be given five index cards with the following compound inequalities written on them:

$$-12 < 4x < 16; \ 6 < 3x - 3 \leq 12; \ -4 < 5x - 2x + 11 < 32; \ -3 \leq -4x + 5 < 21;$$
$$10 < 7x - 8x - 4 < 20$$

They will also be given five index cards with the following solutions written on them:

$$-4 < x \leq 2; \ -3 < x < 4; \ 3 < x \leq 5; \ -24 < x < -14; \ -5 < x < 7$$

Students will work together to match each compound inequality with its solution. They will describe how to add, subtract, multiply, and divide both positive and negative numbers in compound inequalities. They will explain why these types of inequalities are called "compound" inequalities.

Answers

1. $-12 < 4x < 16$ and $-3 < x < 4$

2. $6 < 3x - 3 \leq 12$ and $3 < x \leq 5$

3. $-4 < 5x - 2x + 11 < 32$ and $-5 < x < 7$

4. $-3 \leq -4x + 5 < 21$ and $-4 < x \leq 2$

5. $10 < 7x - 8x - 4 < 20$ and $-24 < x < -14$

6. Add or subtract the number to/from all three sides of the inequality.

7. Multiply or divide the number by all three sides of the inequality.

8. Multiply or divide the negative number by all three sides of the inequality and reverse both inequality signs to their opposite.

9. They are called "compound" inequalities because they are two or more inequalities taken together.

Station 4

Students will be given compound inequalities and a graph of a compound inequality. They will write a compound inequality from a graph. Then they will graph compound inequalities.

Answers

1. $2 \leq x < 8$; I looked at whether there were open or closed circles.

2. Number line:

3. Coefficient = 1; divide each side by 2; $1 \le x < 2$; number line:

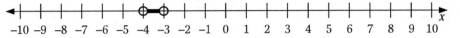

4. Coefficient = 1; divide each side by –10; reverse both signs to their opposite to follow the rule that you reverse the signs when multiplying or dividing in an inequality; $-4 < x < -3$; number line:

Materials List/Setup

Station 1   algebra tiles and slips of paper with the symbols <, >, ≤, and ≥ written on them

Station 2   none

Station 3   five index cards with the following compound inequalities written on them:

$-12 < 4x < 16$; $6 < 3x - 3 \le 12$; $-4 < 5x - 2x + 11 < 32$; $-3 \le -4x + 5 < 21$; $10 < 7x - 8x - 4 < 20$

five index cards with the following solutions written on them:

$-4 < x \le 2$; $-3 < x < 4$; $3 < x \le 5$; $-24 < x < -14$; $-5 < x < 7$

Station 4   none

Discussion Guide

To support students in reflecting on the activities and to gather some formative information about student learning, use the following prompts to facilitate a class discussion to "debrief" the station activities.

Prompts/Questions

1. How is solving an inequality like solving an equation?

2. How is solving an inequality not like solving an equation?

3. What do you do with the inequality signs if you multiply or divide by a negative number?

4. What is a compound inequality?

5. What inequality sign does an open circle represent? Why is the circle not shaded on the number line?

6. What inequality sign does a filled in or closed circle represent? Why is the circle filled in on the number line?

Think, Pair, Share

Have students jot down their own responses to questions, then discuss with a partner (who was not in their station group), and then discuss as a whole class.

Suggested Appropriate Responses

1. You use the order of operations and properties of equality as you would in an equation.

2. In an inequality, you won't just have an equal sign. This means you have more than one solution. Also, in an inequality, you treat multiplying or dividing by a negative coefficient differently.

3. Reverse the inequality signs to its opposite.

4. A compound inequality is two or more inequalities taken together.

5. An open circle represents the < or > inequality signs. The circle is not shaded on the number line to show you that that value is not included in the solution.

6. A closed circle represents the ≤ or ≥ inequality signs. The circle is filled in on the number line to show that that value is included in the solution.

Possible Misunderstandings/Mistakes

- Not reversing the inequality sign to its opposite when multiplying or dividing by a negative coefficient

- Not using the properties of equality for all sides of the inequality (This is especially true for compound inequalities.)

- Using an open circle on a number line for ≤ or ≥

- Using a closed circle on a number line for < or >

- Not checking the solution with appropriate values to make sure the solution is correct

- Not realizing that inequalities give a range of solutions as opposed to just one exact solution

Reasoning with Equations and Inequalities
Set 3: One-Variable Inequalities

Station 1

At this station, you will find algebra tiles and slips of paper with the symbols <, >, ≤, and ≥ written on them. As a group, use these algebra tiles and slips of paper to model each inequality.

1. $2x < 10$

    Use the algebra tiles to solve for x.

    What is the solution of this inequality? _____

    Does the solution $x = 3$ satisfy this inequality? Explain.

    Does the solution $x = 15$ satisfy this inequality? Explain.

    Will the solution of $2x < 10$ have an open or closed circle on the number line? Why or why not?

    Graph this inequality on the number line below.

    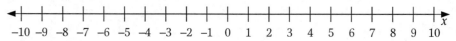

2. $-3x \geq 18$

    Use the algebra tiles to solve for x.

    Is the solution to this inequality $x \geq -6$? Why or why not? Use $x = 5$ to justify your answer.

continued

Reasoning with Equations and Inequalities
Set 3: One-Variable Inequalities

Will the solution of $-3x \geq 18$ have an open or closed circle on the number line? Why or why not?

Graph this inequality on the number line below.

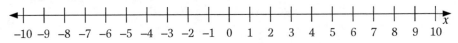

$-10\ -9\ -8\ -7\ -6\ -5\ -4\ -3\ -2\ -1\ 0\ 1\ 2\ 3\ 4\ 5\ 6\ 7\ 8\ 9\ 10$

3. $-\dfrac{1}{4}x < 2$

Is the solution to this inequality $x < -8$? Why or why not? Use $x = -12$ to justify your answer.

Will the solution of $-\dfrac{1}{4}x < 2$ have an open or closed circle on the number line? Why or why not?

Graph this inequality on the number line below.

$-10\ -9\ -8\ -7\ -6\ -5\ -4\ -3\ -2\ -1\ 0\ 1\ 2\ 3\ 4\ 5\ 6\ 7\ 8\ 9\ 10$

4. Based on your answers in problems 2 and 3, derive a rule for multiplication and division by a negative number when solving an inequality. Write your rule below.

Reasoning with Equations and Inequalities
Set 3: One-Variable Inequalities

Station 2

Work as a group to solve and graph these inequalities.

1. $4x - 8 \geq 20$

   What is the first step you take to solve this inequality?

   What is the second step you take to solve this inequality?

   What is the solution of this inequality? _____

   How can you double-check your solution?

   Will the solution of $4x - 8 \geq 20$ have an open or closed circle on the number line? Why or why not?

   Graph this inequality on the number line below.

2. $-10x - 19 \leq 31$

   What is the first step you take to solve this inequality?

   What is the second step you take to solve this inequality?

continued

Reasoning with Equations and Inequalities
Set 3: One-Variable Inequalities

What happens to the sign of an inequality when you multiply or divide by a negative number?

What is the solution of this inequality? _____

How can you double-check your solution?

Will the solution of $-10x - 19 \leq 31$ have an open or closed circle on the number line? Why or why not?

Graph this inequality on the number line below.

3. $-\dfrac{x}{5} > 15 + \dfrac{2}{5}x$

What is the first step you take to solve this inequality?

What is the second step you take to solve this inequality?

What happens to the sign of an inequality when you multiply or divide by a negative number?

What is the solution of this inequality? _____

How can you double-check your solution?

continued

Will the solution of $-\dfrac{x}{5} > 15 + \dfrac{2}{5}x$ have an open or closed circle on the number line? Why or why not?

Graph this inequality on the number line below.

Algebra I Station Activities for Common Core State Standards

Reasoning with Equations and Inequalities
Set 3: One-Variable Inequalities

Station 3

At this station, you will find five index cards with the following compound inequalities written on them:

$$-12 < 4x < 16; \; 6 < 3x - 3 \le 12; \; -4 < 5x - 2x + 11 < 32; \; -3 \le -4x + 5 < 21; \; 10 < 7x - 8x - 4 < 20$$

You will also find five index cards with the following solutions written on them:

$$-4 < x \le 2; \; -3 < x < 4; \; 3 < x \le 5; \; -24 < x < -14; \; -5 < x < 7$$

Place the index cards in one pile and shuffle. Then work as a group to match each compound inequality with its solution. Write your answers below.

1. _____

2. _____

3. _____

4. _____

5. _____

6. How did you add or subtract a constant in order to find the solution?

7. How did you multiply or divide a positive coefficient of the variable in order to find the solution?

8. How did you multiply or divide a negative coefficient of the variable in order to find the solution? What extra step did you take with the inequality signs?

9. Why do you think these inequalities are called "compound" inequalities?

Reasoning with Equations and Inequalities
Set 3: One-Variable Inequalities

Station 4

You will be given compound inequalities or graphs of compound inequalities. Work together to answer the questions.

1. A compound inequality is given by this graph:

    Write a compound inequality for this graph using the variable x. _____

    How did you determine the signs (<, >, ≤, or ≥) you used in the compound inequality?

2. On the number line below, graph the compound inequality $-6 < x \leq 2$.

3. Use the compound inequality $2 \leq 2x < 4$ to answer the questions.

    What must the coefficient of x be before you can graph this inequality?

    _____

    What was the first step you took to solve for x in the inequality above?

    Rewrite the inequality: _____

    Graph the inequality on the number line below.

continued

Reasoning with Equations and Inequalities
Set 3: One-Variable Inequalities

4. Use the compound inequality $30 < -10x < 40$ to answer the questions.

   What must the coefficient of x be before you can graph this inequality?

   _____

   What was the first step you took to solve for x in the inequality above?

   What do you have to do to the signs ($<$, $>$, $\leq$, or $\geq$) in the inequality? Why?

   Rewrite the inequality: _____

   Graph the inequality on the number line below.

Reasoning with Equations and Inequalities

Set 4: Two-Variable Inequalities

Goal: To provide opportunities for students to develop concepts and skills related to solving and graphing two-variable linear inequalities

Common Core Standards

Algebra: Reasoning with Equations and Inequalities

Represent and solve equations and inequalities graphically.

A-REI.12. Graph the solutions to a linear inequality in two variables as a half-plane (excluding the boundary in the case of a strict inequality), and graph the solution set to a system of linear inequalities in two variables as the intersection of the corresponding half-planes.

Student Activities Overview and Answer Key

Station 1

Students will be given 10 index cards with the following numbers written on them: −8, −4, −2, 0, 1, 2, 4, 8, 10, and 15. They will use the index cards to populate inequalities with two variables. They will create ordered pairs and determine whether or not the ordered pairs are solutions to the inequalities.

Answers

1. Answers will vary; verify that students have written the (x, y) ordered pair correctly. Verify they correctly substituted the x- and y-values into the inequality to determine whether or not it was a solution.

2. Answers will vary; verify that students have written the (x, y) ordered pair correctly. Verify that students correctly substituted the x- and y-values into the inequality to determine whether or not it was a solution.

3. Answers will vary; verify that students have written the (x, y) ordered pair correctly. Verify that students correctly substituted the x- and y-values into the inequality to determine whether or not it was a solution.

4. Answers will vary; verify that students have written the (x, y) ordered pair correctly. Verify that students correctly substituted the x- and y-values into the inequality to determine whether or not it was a solution.

5. x and y are in an ordered pair (x, y). Switching their values may yield solutions or no solution. It depends on the inequality and the values of the x- and y-values.

Station 2

Students will be given graph paper and a ruler. Students will find the slope, *x*-intercept, and *y*-intercept of a two-variable inequality. They will graph the inequality. They will describe why the line is dashed and why the graph is shaded above the line.

Answers

1. $4x + 2y > 6$; $2y > -4x + 6$; $y > -2x + 3$

2. $m = -2$

3. $(0, 3)$

4. $(3/2, 0)$

5. It is a dashed line because of the > sign, so the ordered pairs on the line are not solutions.

6. No, because of the > sign. The points are above the line.

7. Test ordered pairs above and below the line to see if they are solutions.

8. $(6, 10)$

9. above the line

10.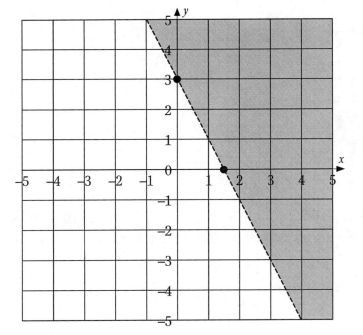

Station 3

Students will be given a graphing calculator. They will graph a two-variable inequality given step-by-step instructions. They find the type of line and the *x*- and *y*-intercepts of the graph. Then they graph a two-variable inequality by themselves. They will also analyze the graph.

Answers

1. It is a solid line because of the $\geq$ sign.

2. $(0, 5), (-5/2, 0)$

3. Below the line because of the < sign.

4.

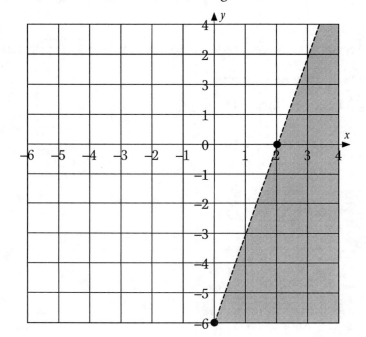

Station 4

Students will be given graph paper, a ruler, and a two-variable inequality. They will complete a table of ordered pairs to see if the ordered pairs satisfy the inequality. Then students will describe and graph the inequality.

Answers

1.

Ordered pair	x	y	$10x + 9$	Is $y \le 10x + 9$?
(2, 3)	2	3	29	yes
(0, 9)	0	9	9	yes
(−10, 4)	−10	4	−91	no
(5, −8)	5	−8	59	yes

2. (0, 9); y-intercept

3. (−9/10, 0)

4. Shade to the right because the ordered pairs to the right of the line satisfy the inequality.

5.

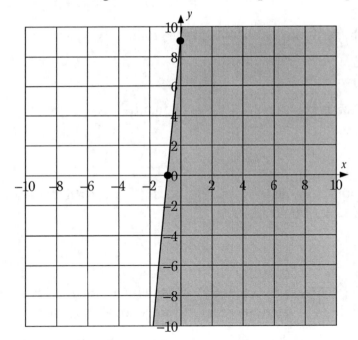

Materials List/Setup

Station 1 10 index cards with the following numbers written on them:

−8, −4, −2, 0, 1, 2, 4, 8, 10, 15

Station 2 graph paper; ruler

Station 3 graphing calculator

Station 4 graph paper; ruler

Discussion Guide

To support students in reflecting on the activities and to gather some formative information about student learning, use the following prompts to facilitate a class discussion to "debrief" the station activities.

Prompts/Questions

1. How can you use ordered pairs to find solutions to a two-variable linear inequality?

2. How do you know if the graph of a two-variable linear inequality should have a solid or dashed line?

3. How can you determine whether to shade above or below the line in the graph of a two-variable linear inequality?

4. For what inequality sign(s) are the solutions on the line in the graph of two-variable linear inequalities?

5. How can you find the slope of a two-variable inequality?

6. How can you find the x- and y-intercepts of a two-variable inequality?

Think, Pair, Share

Have students jot down their own responses to questions, then discuss with a partner (who was not in their station group), and then discuss as a whole class.

Suggested Appropriate Responses

1. Plug the ordered pair into the inequality. If this makes a true statement, then the ordered pair is a solution. If this makes a false statement, then the ordered pair is not a solution.

2. Solid lines are for $\leq$ and $\geq$ inequality signs. Dashed lines are for $<$ and $>$ signs.

3. Check ordered pairs above and below the line to see which ordered pair is a solution. Shade the area with ordered pairs that are solutions.

4. $\leq$ and $\geq$ inequality signs have solutions on the line.

5. Write the inequality in $y = mx + b$ form. The slope is represented by m.

6. Write the equation in $y = mx + b$ form. Find the x-intercept by setting $y = 0$ and solving for x. Find the y-intercept by setting $x = 0$ and solving for y.

Possible Misunderstandings/Mistakes

- Not recognizing that the line is dashed for < and > signs
- Not recognizing that the line is solid for ≤ and ≥ signs
- Incorrectly substituting the ordered pair into the equation to see if it is a solution to the inequality
- Not shading either side of the graph to show all the possible solutions
- Not understanding that you should use points above and below the graph to see which side of the graph needs to be shaded

Reasoning with Equations and Inequalities
Set 4: Two-Variable Inequalities

Station 1

At this station, you will find 10 index cards with the following numbers written on them:

$$-8, -4, -2, 0, 1, 2, 4, 8, 10, 15$$

Place the index cards in a pile. As a group, draw one card from the pile and write the result on the first line below. Repeat this process for the second line. Both lines should now contain a number (one number for x, one number for y). Write the numbers as an ordered pair, then answer the question. Repeat this process for each problem that follows.

1. You are given an inequality with two variables: $2x + y > 4$

   As a group, draw one card from the pile and write it as $x =$ _____

   Draw another card from the pile and write it as $y =$ _____

   Write these values as an ordered pair. _____

   Is this ordered pair a solution to the inequality? Why or why not?

2. You are given an inequality with two variables: $3x - 4y < 40$

   As a group, draw one card from the pile and write it as $x =$ _____

   Draw another card from the pile and write it as $y =$ _____

   Write these values as an ordered pair. _____

   Is this ordered pair a solution to the inequality? Why or why not?

3. You are given an inequality with two variables: $x + 9y \leq 8$

   As a group, draw one card from the pile and write it as $x =$ _____

   Draw another card from the pile and write it as $y =$ _____

   Write these values as an ordered pair. _____

   Is this ordered pair a solution to the inequality? Why or why not?

continued

Algebra I Station Activities for Common Core State Standards

Reasoning with Equations and Inequalities
Set 4: Two-Variable Inequalities

4. You are given an inequality with two variables: $-2x - \dfrac{1}{2}y > 0$

    As a group, draw one card from the pile and write it as $x =$ _____

    Draw another card from the pile and write it as $y =$ _____

    Write these values as an ordered pair. _____

    Is this ordered pair a solution to the inequality? Why or why not?

5. How would switching the x- and y-values for problems 1–4 affect your answers? Explain your reasoning.

Reasoning with Equations and Inequalities
Set 4: Two-Variable Inequalities

Station 2

At this station, you will find graph paper and a ruler. Work together to answer the questions and graph the inequalities.

1. Solve the inequality $4x + 2y > 6$ for y. Show your work.

2. What is the slope of the inequality? _____

3. What is the y-intercept of the inequality? _____

4. What is the x-intercept of the inequality? _____

5. Use the x- and y-intercepts to graph this inequality on your graph paper. Draw a straight dashed line to construct the graph.

    Why do you think the line should be dashed instead of solid?

6. Does the line on your graph show the solutions to the inequality? Are there more solutions? Explain your answer.

7. How can you figure out whether the solutions to the inequality lie below or above the line on your graph?

continued

Reasoning with Equations and Inequalities
Set 4: Two-Variable Inequalities

8. Which ordered pair satisfies $4x + 2y > 6$: (6, 10) or (–6, –10)?

9. Based on your answer to problem 8, should you shade above or below the line on your graph?

10. On your graph paper, shade the area you specified in problem 9.

Reasoning with Equations and Inequalities
Set 4: Two-Variable Inequalities

Station 3

At this station, you will find a graphing calculator. Follow the steps described below to graph a two-variable inequality on the graphing calculator.

Inequality: $y \geq 2x + 5$

> **Steps for graphing the inequality on your graphing calculator:**
>
> 1. At the Y1 prompt, type in "$2x + 5$".
>
> 2. Use the left arrow to go to the left hand side of the "Y1". Then hit "Enter" until the shaded above symbol, , is displayed. Use the shaded above symbol because of the $\geq$ sign.
>
> 3. Hit the "ZOOM" key. Select "#6 Zstandard".
>
> 4. Hit the "GRAPH" key.

Now that you've graphed the inequality, use the graph to answer the following questions.

1. Graphing calculators will always show a solid line. For this inequality, should the line be solid or dashed? Explain your answer.

2. You can use the "Trace" key to find the x- and y-intercepts of inequalities. What are the x- and y-intercepts of the inequality you graphed?

3. Use your graphing calculator to graph $y < 3x - 6$.

    Should you shade above or below the line? Explain your answer.

continued

Reasoning with Equations and Inequalities

Set 4: Two-Variable Inequalities

4. Draw the graph you created in the space below.

Reasoning with Equations and Inequalities
Set 4: Two-Variable Inequalities

Station 4

At this station, you will find graph paper and a ruler. Use the two-variable inequality $y \leq 10x + 9$ to answer the problems.

1. Work together to fill in the table below.

Ordered pair	x	y	$10x + 9$	Is $y \leq 10x + 9$?
(2, 3)				
(0, 9)				
	−10	4		
	5	−8		

2. Which ordered pair is on the line of the inequality? _____

   Is this the x- or y-intercept? _____

3. Find the other intercept to give you two points on the line of the inequality. Write this intercept on the line below.

   _____

   As a group, graph the line of the inequality on your graph paper.

4. The other ordered pairs in the table tell you where to shade the graph of the inequality. Based on the table, should you shade to the right or left of the line? Explain your answer.

5. Shade the appropriate side of your graph of the inequality.

Reasoning with Equations and Inequalities

Set 5: Solving 2-by-2 Systems by Graphing

Goal: To provide opportunities for students to develop concepts and skills related to solving linear systems of equations by graphing

Common Core Standards

Algebra: Reasoning with Equations and Inequalities

Solve systems of equations.

A-REI.6. Solve systems of linear equations exactly and approximately (e.g., with graphs), focusing on pairs of linear equations in two variables.

Represent and solve equations and inequalities graphically.

A-REI.11. Explain why the x-coordinates of the points where the graphs of the equations $y = f(x)$ and $y = g(x)$ intersect are the solutions of the equation $f(x) = g(x)$; find the solutions approximately; e.g., using technology to graph the functions, make tables of values, or find successive approximations. Include cases where $f(x)$ and/or $g(x)$ are linear, polynomial, rational, absolute value, exponential, and logarithmic functions.★

Student Activities Overview and Answer Key

Station 1

Students will be given graph paper and a ruler. Students will graph a system of linear equations. They will find the point of intersection of the lines and realize that this is the solution to the linear system. They will double-check the point of intersection by substituting it into the original equations.

Answers

1.

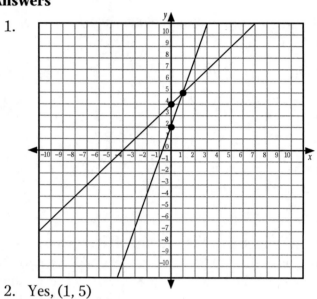

2. Yes, (1, 5)

3. The point of intersection is the solution of the system of linear equations.

4. Substitute (1, 5) into both equations to verify that it satisfies both equations.

5.

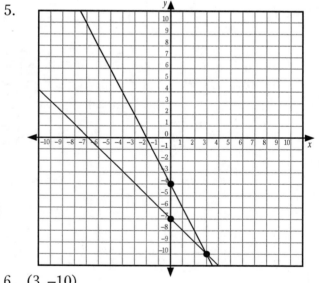

6. (3, −10)

Station 2

Students will be given four index cards with the following systems of linear equations written on them:

$$\begin{cases} 10x - 2y = 10 \\ 5x - y = 5 \end{cases}; \quad \begin{cases} 5x + y = -3 \\ 7x + y = -5 \end{cases}; \quad \begin{cases} y = 4 \\ -x + y = 10 \end{cases}; \quad \begin{cases} y = 2x + 5 \\ -2x + y = 8 \end{cases}$$

Students will work together to match each system of linear equations with the appropriate graph. They will explain the strategy they used to match the graphs.

Answers

1. $\begin{cases} 5x + y = -3 \\ 7x + y = -5 \end{cases}$

2. $\begin{cases} y = 4 \\ -x + y = 10 \end{cases}$

3. $\begin{cases} 10x - 2y = 10 \\ 5x - y = 5 \end{cases}$

4. $\begin{cases} y = 2x + 5 \\ -2x + y = 8 \end{cases}$

5. Answers will vary. Possible answer: Finding the point of intersection of the lines. This is the solution to the linear system. Substitute this point into the equations to see which systems of linear equations have this solution.

Station 3

Students will be given linear equations and will find points that satisfy the equations. Then they will look at the equations as a linear system. They will find the solution, or no solution, of the linear system. They will describe the strategy they used to determine the solution of the linear system.

Answers

1.

x	y
−5	−14
0	−4
5	6
10	16
15	26

2.

x	y
−4	37
0	31
8	19
10	16
12	13

3. Yes, (10, 16)

4. They had the same ordered pair of (10, 16), which means the lines intersect at that point.

5.

x	y
−2	6
0	10
2	14
4	18

6.

x	y
−2	−14
0	−10
2	−6
4	−2

7. The table of values may or may not contain the solution.

8. You can use the slope of each line and y-intercepts, or you can graph the equations and see if they intersect.

9. There is no solution. Each equation has the same slope, but different y-intercepts.

Station 4

Students will be given a graphing calculator. Students will be shown how to find the solution of a system of linear equations using the graphing calculator. Then they will use the graphing calculator to find the solutions of given systems of linear equations.

Answers

1. (−1.67, −9.67)

2. (−8, −12)

3. (1.29, 3.86)

4. (0.73, 3.27)

Materials List/Setup

Station 1   graph paper; ruler

Station 2   four index cards with the following systems of linear equations written on them:

$$\begin{cases} 10x - 2y = 10 \\ 5x - y = 5 \end{cases} ; \begin{cases} 5x + y = -3 \\ 7x + y = -5 \end{cases} ; \begin{cases} y = 4 \\ -x + y = 10 \end{cases} ; \begin{cases} y = 2x + 5 \\ -2x + y = 8 \end{cases}$$

Station 3   none

Station 4   graphing calculator

Discussion Guide

To support students in reflecting on the activities and to gather some formative information about student learning, use the following prompts to facilitate a class discussion to "debrief" the station activities.

Prompts/Questions

1. How do you solve systems of linear equations by graphing?

2. How do you know if a system of linear equations has infinite solutions?

3. How do you know if a system of linear equations has no solutions?

4. How can you use a graphing calculator to graph and solve systems of linear equations?

5. Give an example of a graph of a linear system that is used in the real world.

Think, Pair, Share

Have students jot down their own responses to questions, then discuss with a partner (who was not in their station group), and then discuss as a whole class.

Suggested Appropriate Responses

1. Graph each line. Their point of intersection is the solution to the linear system.

2. Graph each line. If the lines are the same line then they have infinite solutions. Or, if the lines have the same slope and y-intercept, then there are infinite solutions.

3. Graph each line. If the lines are parallel or have no point of intersection, the system has no solutions.

4. Type the equations into Y1 and Y2. Hit "GRAPH", "2nd", "Trace", and select "5: Intersect". Then hit the "ENTER" key three times to find the point of intersection. The point of intersection is the solution to the linear system.

5. Answers will vary. Possible answer: comparing service plans from two different cell phone companies

Possible Misunderstandings/Mistakes

- Incorrectly graphing the equations by not finding the correct slope and intercepts

- Not realizing that equations with the same slope and y-intercept are the same line. If they are a system of linear equations then the system has infinite solutions.

- Not realizing that parallel lines have no solutions

- Not realizing the point of intersection of the two lines is the solution to the system of linear equations

Reasoning with Equations and Inequalities
Set 5: Solving 2-by-2 Systems by Graphing

Station 1

At this station, you will find graph paper and a ruler. You can find the solution to a system of linear equations by graphing each equation in the same coordinate plane.

1. As a group, graph each of the following linear equations on the same graph.

    $y = 3x + 2$

    $y = x + 4$

2. Do the lines intersect? If so, find the point at which the two lines intersect.

3. What does this point of intersection tell you about the two equations?

4. How can you double-check your answer to problem 3?

5. As a group, graph the system of linear equations on a new graph.
    $$\begin{cases} y = -2x - 4 \\ y = -x - 7 \end{cases}$$

6. Using this graph, what is the solution to the linear system?

Reasoning with Equations and Inequalities
Set 5: Solving 2-by-2 Systems by Graphing

Station 2

At this station, you will find four index cards with the following linear systems of equations written on them:

$$\begin{cases} 10x - 2y = 10 \\ 5x - y = 5 \end{cases}; \begin{cases} 5x + y = -3 \\ 7x + y = -5 \end{cases}; \begin{cases} y = 4 \\ -x + y = 10 \end{cases}; \begin{cases} y = 2x + 5 \\ -2x + y = 8 \end{cases}$$

Work together to match each system of linear equations with the appropriate graph below. Write the appropriate system of linear equations under each graph.

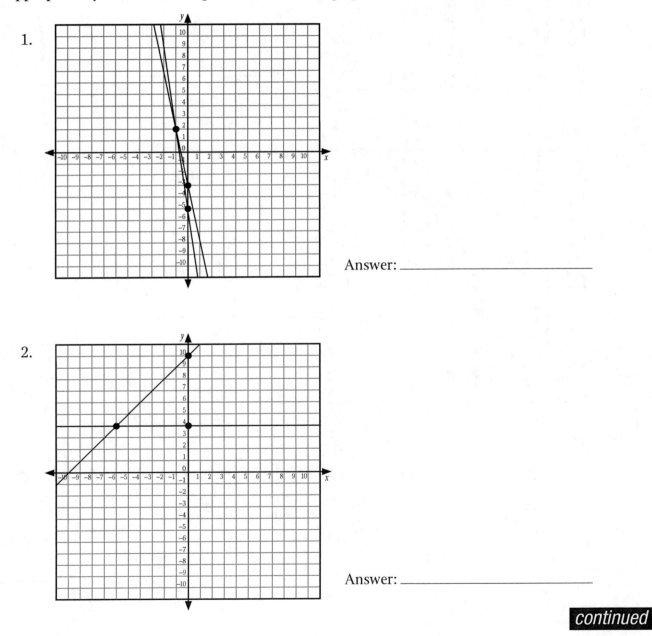

1.

Answer: _____

2.

Answer: _____

continued

Reasoning with Equations and Inequalities
Set 5: Solving 2-by-2 Systems by Graphing

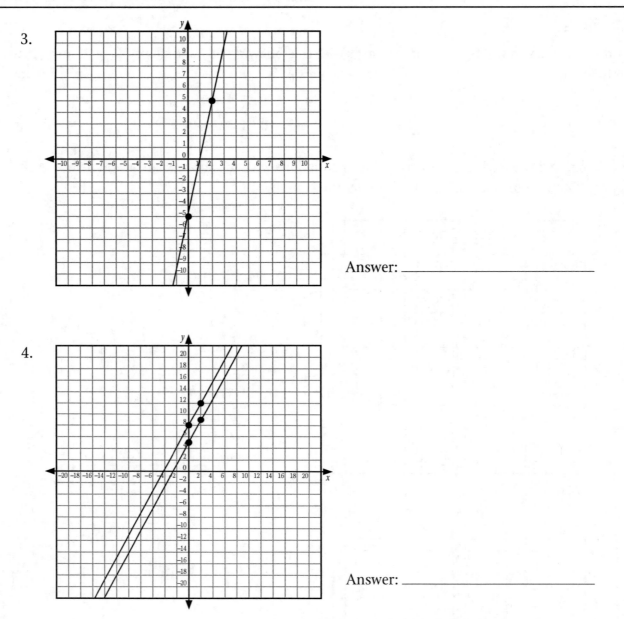

3.

Answer: _____

4.

Answer: _____

5. What strategy did you use to match the systems of linear equations with the appropriate graph?

Reasoning with Equations and Inequalities
Set 5: Solving 2-by-2 Systems by Graphing

Station 3

Use what you know about linear equations to solve the following problems.

1. Complete the table below for the linear equation $2x - y = 4$.

x	y
-5	
0	
5	
10	
15	

2. Complete the table below for the linear equation $3x + 2y = 62$.

x	y
-4	
0	
8	
10	
12	

3. If $\begin{cases} 2x - y = 4 \\ 3x + 2y = 62 \end{cases}$ were a set of linear equations, would it have a solution? If so, what would it be?

4. What strategy did you use in problem 3 to find your answer?

5. Complete the table below for the linear equation $2x = y - 10$.

x	y
-2	
0	
2	
4	

continued

Reasoning with Equations and Inequalities

Set 5: Solving 2-by-2 Systems by Graphing

6. Complete the table below for the linear equation $2x = y + 10$.

x	y
−2	
0	
2	
4	

7. If $\begin{cases} 2x = y - 10 \\ 2x = y + 10 \end{cases}$ were a set of linear equations, why would you need more information than just the values in the tables you found for problems 5 and 6 to determine any solutions?

8. What information would you need to determine whether or not a system of linear equations has a solution, infinite solutions, or no solutions?

9. Does $\begin{cases} 2x = y - 10 \\ 2x = y + 10 \end{cases}$ have a solution? Why or why not?

Reasoning with Equations and Inequalities
Set 5: Solving 2-by-2 Systems by Graphing

Station 4

At this station, you will find a graphing calculator. You can graph systems of linear equations on your calculator to find their point of intersection. Work together to graph systems of linear equations on your calculator.

Given: $\begin{cases} y = 4x - 3 \\ y = x - 8 \end{cases}$

Follow these steps on your graphing calculator to find the point of intersection for this system of equations:

1. Hit the "Y =" key. Type Y1 = $4x - 3$.

2. Type Y2 = $x - 8$.

3. Hit the "GRAPH" key.

4. Hit "2nd" and then "TRACE". Select "5: Intersect".

5. Hit "ENTER" three times. You will be given the x- and y-values of the intersection.

---

1. What is the intersection/solution of $\begin{cases} y = 4x - 3 \\ y = x - 8 \end{cases}$? _____

Using the same steps, find the solution to each system of linear equations below on your graphing calculator.

2. $\begin{cases} y = 3x + 12 \\ y = x - 4 \end{cases}$

   Solution = _____

3. $\begin{cases} y = -4x + 9 \\ y = 3x \end{cases}$

   Solution = _____

4. $\begin{cases} 2x + 2y = 8 \\ y = 10x - 4 \end{cases}$

   Solution = _____

Reasoning with Equations and Inequalities

Instruction

Goal: To provide opportunities for students to develop concepts and skills related to solving systems of linear equations using substitution

Common Core Standards

Algebra: Reasoning with Equations and Inequalities

Solve systems of equations.

A-REI.6. Solve systems of linear equations exactly and approximately (e.g., with graphs), focusing on pairs of linear equations in two variables.

Student Activities Overview and Answer Key

Station 1

Students will be given 10 blue algebra tiles to represent the coefficient of x, 10 yellow algebra tiles to represent the coefficient of y, and 40 red algebra tiles to represent the constant. They will also be given index cards with x, y, +, −, and = written on them.

Students use the algebra tiles and index cards to model a system of linear equations. They rearrange the tiles and index cards to solve for x. Then they solve for y to find the solution to the system of linear equations.

Answers

1. x and y

2. $x = -y + 10$

3. $2(-y + 10) + y = 15; y = 5$

4. Substitute y into the first equation to find $x = 5$.

5. $x + y = 10$ and $5 + 5 = 10$; $2x + y = 15$ and $2(5) + 5 = 15$

Station 2

Students will be given systems of linear equations. They will work together to solve each system of linear equations using the substitution method. They will show their work and explain how to double-check their answer.

Answers

1. $x = y + 3$

    $5(y + 3) - y = -5$

    $5y + 15 - y = -5$

    $y = -5$

    and

    $5x - (-5) = -5$

    $x = -2$

2. For $5x = 4y + 1$, $x = \dfrac{4}{5}y + \dfrac{1}{5}$, so $7\left(\dfrac{4}{5}y + \dfrac{1}{5}\right) + 4y = 11$, $y = 1$ and for

    $5x - 4(1) = 1$, $x = 1$.

3. Solve for one variable in the first equation. Substitute this quantity in for that variable in the second equation. Solve for the variable still in the equation. Then substitute your answer for that variable into the original equation to find the value of the other variable.

4. Substitute x and y into both equations to make sure they satisfy both equations.

5. You substitute one variable in terms of the other variable to find x or y.

Station 3

Students will be given a system of linear equations and nine index cards with the following equations written on them:

$y = 1$; $10x = 30$; $y = 2x - 5$; $x = 3$; $2x - 12x + 30 = 0$; $-10x + 30 = 0$;

$2(3) - y = 5$; $2x - 6(2x - 5) = 0$; and $6 - y = 5$

Each index card represents a step in solving a system of linear equations by substitution. Students work together to arrange the steps in the correct order. They give the strategy they used to arrange the index cards.

Answers

1. $y = 2x - 5$

2. $2x - 6(2x - 5) = 0$

3. $2x - 12x + 30 = 0$

4. $-10x + 30 = 0$

5. $10x = 30$

6. $x = 3$

7. $2(3) - y = 5$

8. $6 - y = 5$

9. $y = 1$

10. Answers will vary. Possible answer: I looked for one variable written in terms of the other variable. Then I solved by substitution in a step-by-step manner.

Station 4

Students will be given systems of linear equations that have either infinite solutions or no solutions. They will try to find the solution of the system. Then they will realize that lines with the same slope and y-intercept have infinite solutions. They will also find that lines that have the same slope but different y-intercepts have no solutions.

Answers

1. $y = 4 - x$
   $2x + 2(4 - x) = 8$
   $8 = 8$

2. Answers will vary.

3. -1

4. -1

5. yes; yes

6. When lines are parallel and have the same y-intercept, they are the same line. This means the system of linear equations has an infinite number of solutions.

7. $4 \neq 10$

8. no solutions

9. 2

10. 2

11. yes

12. no

13. The lines have the same slope, but different y-intercepts. There are no solutions to the system of linear equations.

Materials List/Setup

Station 1   10 blue algebra tiles to represent the coefficient of x, 10 yellow algebra tiles to represent the coefficient of y, and 40 red algebra tiles to represent the constant; index cards with x, y, +, −, and = written on them

Station 2   none

Station 3   nine index cards with the following written on them:

$y = 1$; $10x = 30$; $y = 2x - 5$; $x = 3$; $2x - 12x + 30 = 0$; $-10x + 30 = 0$;

$2(3) - y = 5$; $2x - 6(2x - 5) = 0$; and $6 - y = 5$

Station 4   none

Discussion Guide

To support students in reflecting on the activities and to gather some formative information about student learning, use the following prompts to facilitate a class discussion to "debrief" the station activities.

Prompts/Questions

1. How do you solve systems of linear equations using substitution?

2. How many solutions does a system of linear equations have if the equations have the same slope and *y*-intercept?

3. How many solutions does a system of linear equations have if the equations have the same slope, but different *y*-intercepts?

4. Why does the "substitution method" have this name?

Think, Pair, Share

Have students jot down their own responses to questions, then discuss with a partner (who was not in their station group), and then discuss as a whole class.

Suggested Appropriate Responses

1. Solve for one variable in the first equation. Substitute this quantity in for that variable in the second equation. Solve for the variable still in the equation. Then substitute your answer for that variable into the original equation to find the value of the other variable.

2. infinite number of solutions

3. no solutions

4. You substitute one variable in terms of the other variable to find *x* or *y*.

Possible Misunderstandings/Mistakes

- Incorrectly solving for one variable when rewriting an equation in step 1 of the process

- Incorrectly substituting *x* in the first equation for *y* in the second equation

- Not double-checking that the values of *x* and *y* satisfy both equations

- Not realizing that lines that have the same slope and *y*-intercept are actually the same line. These have an infinite number of solutions to the system of linear equations.

- Not realizing that parallel lines that have different *y*-intercepts will have no solutions for the system of linear equations

Reasoning with Equations and Inequalities
Set 6: Solving 2-by-2 Systems by Substitution

Station 1

At this station, you will find 10 blue algebra tiles to represent the coefficient of x, 10 yellow algebra tiles to represent the coefficient of y, and 40 red algebra tiles to represent the constant. You will also be given index cards with x, y, $+$, $-$, and $=$ written on them.

As a group, arrange the algebra tiles and index cards to model the system of linear equations below:

$$\begin{cases} x + y = 10 \\ 2x + y = 15 \end{cases}$$

1. What values do you need to find in order to solve this system of linear equations?

2. Rearrange $x + y = 10$ to solve for x. Write your answer on the line below.

    _____

3. Substitute what you found in problem 2 into the equation $2x + y = 15$. Show your work in the space below.

    Now solve for y. What is the value of y?

    _____

4. How can you use this value of y to find x?

    What is the value of x?

    _____

5. Verify that x and y satisfy both equations. Show your work in the space below.

Reasoning with Equations and Inequalities
Set 6: Solving 2-by-2 Systems by Substitution

Station 2

Use substitution to find the solutions for each system of linear equations. Show your work.

1. $\begin{cases} 5x - y = -5 \\ x - y = 3 \end{cases}$

2. $\begin{cases} 5x - 4y = 1 \\ 7x + 4y = 11 \end{cases}$

3. What strategies did you use to find x and y?

4. How can you double-check your solutions for x and y?

5. Why is this method called the "substitution" method?

Reasoning with Equations and Inequalities
Set 6: Solving 2-by-2 Systems by Substitution

Station 3

At this station, you will find nine index cards with the following written on them:

$y = 1$; $10x = 30$; $y = 2x - 5$; $x = 3$; $2x - 12x + 30 = 0$; $-10x + 30 = 0$;

$2(3) - y = 5$; $2x - 6(2x - 5) = 0$; and $6 - y = 5$

Given: $\begin{cases} 2x - y = 5 \\ 2x - 6y = 0 \end{cases}$

   Each index card represents one step in solving the linear system of equations by substitution. Shuffle the index cards. Work together to arrange the index cards in the correct order of solving the system of linear equations by substitution.

List the steps below.

1. _____

2. _____

3. _____

4. _____

5. _____

6. _____

7. _____

8. _____

9. _____

10. What strategy did you use to find the order of the steps?

Reasoning with Equations and Inequalities
Set 6: Solving 2-by-2 Systems by Substitution

Station 4

Work as a group to solve this system of linear equations by substitution.

$$\begin{cases} x + y = 4 \\ 2x + 2y = 8 \end{cases}$$

1. What happens when you solve this equation by substitution?

2. Based on problem 1, how many solutions do you think this system of linear equations has?

   _____

3. What is the slope of $x + y = 4$? _____

4. What is the slope of $2x + 2y = 8$? _____

5. Are these two lines parallel? _____

   Do these lines have the same y-intercept? _____

6. How can you relate the slopes and y-intercepts of each equation to the number of solutions for the linear system of equations?

continued

Reasoning with Equations and Inequalities
Set 6: Solving 2-by-2 Systems by Substitution

Work as a group to solve this system of linear equations by substitution.

$$\begin{cases} y = 2x + 4 \\ y = 2x + 10 \end{cases}$$

7. What happens when you solve this equation by substitution?

8. Based on problem 7, how many solutions do you think this system of linear equations has?

    _____

9. What is the slope of $y = 2x + 4$? _____

10. What is the slope of $y = 2x + 10$? _____

11. Are these two lines parallel? _____

12. Do these lines have the same y-intercept? _____

13. How can you relate the slopes and y-intercepts of each equation to the number of solutions for the linear system of equations?

Reasoning with Equations and Inequalities

Set 7: Solving 2-by-2 Systems by Elimination

Goal: To provide opportunities for students to develop concepts and skills related to solving systems of linear equations using multiplication and addition

Common Core Standards

Algebra: Reasoning with Equations and Inequalities

Solve systems of equations.

> **A-REI.5.** Prove that, given a system of two equations in two variables, replacing one equation by the sum of that equation and a multiple of the other produces a system with the same solutions.

Student Activities Overview and Answer Key

Station 1

Students will be given an equation. Students will use the addition property to solve the equation. Then they will apply the addition property to a system of linear equations.

Answers

1. Combine like terms $(x + x)$. Subtract 6 from both sides or add -6 to both sides of the equation using the addition property.

2. $x = 4$

3. y because you have $-1y + 1y = 0y$

4. $2x = 16$, so $x = 8$

5. $y = 3$

6. $(8, 3)$

7. $3y = 24$, so $y = 8$; $x = 1$; $(1, 8)$

8. $-2y = 18$, so $y = -9$; $x = -1$; $(-1, -9)$

Station 2

Students will be given 10 yellow, 10 red, and 20 green algebra tiles. Students are also given index cards with the symbols x, y, $-$, $+$, and $=$ written on them. They will use the algebra tiles and index cards to model and solve systems of linear equations by using the addition/elimination method.

Answers

1. yellow algebra tiles

2. red algebra tiles

3. Student drawings should show 7 yellow tiles ($7x$) = 14 green tiles.

4. $x = 2$

5. Substitute x into the first equation to solve for y.

6. $(2, 1)$

7. $(2, 16)$

Station 3

Students will be given a system of linear equations. They solve the system using multiplication and addition on the x variable in the elimination method. Then they solve the same systems using multiplication and addition on the y variable in the elimination method.

Answers

1. No, because the same variable must have opposite coefficients in the two equations.

2. -2

3. $4x + 10y = 42$

   $-4x - 2y = -10$

   $8y = 32$

   $y = 4$

4. $4x + 10(4) = 42$

   $x = \frac{1}{2}$; solution is $(\frac{1}{2}, 4)$

5. $4x + 10y = 42$

   $-20x - 10y = -50$

   $x = \frac{1}{2}$, so $y = 4$; solution is $(\frac{1}{2}, 4)$

Station 4

Students will be given eight index cards with the following written on them:

$$5x = 0; \quad \begin{cases} 10x + 4y = 16 \\ -10x - 15y = -60 \end{cases}; (0, 4); \quad \begin{cases} 2(5x + 2y) = 2(8) \\ -5(2x + 3y) = -5(12) \end{cases}; x = 0; -11y = -44;$$

$$5x + 2(4) = 8; y = 4$$

Students work together to arrange the index cards to reflect how to find the solution of the system of linear equations. Then they explain the strategy they used to find this order.

Answers

1. $\begin{cases} 2(5x + 2y) = 2(8) \\ -5(2x + 3y) = -5(12) \end{cases}$

2. $\begin{cases} 10x + 4y = 16 \\ -10x - 15y = -60 \end{cases}$

3. $-11y = -44$

4. $y = 4$

5. $5x + 2(4) = 8$

6. $5x = 0$

7. $x = 0$

8. $(0, 4)$

9. Answers will vary. Possible answers: We figured out which variable was going to be multiplied and added to cancel it out. Then we figured out the rest of the steps from this information.

Materials List/Setup

Station 1   none

Station 2   10 yellow, 10 red, and 20 green algebra tiles; index cards with the symbols x, y, $-$, $+$, and $=$ written on them

Station 3   none

Station 4   eight index cards with the following written on them:

$$5x = 0; \quad \begin{cases} 10x + 4y = 16 \\ -10x - 15y = -60 \end{cases}; (0, 4); \quad \begin{cases} 2(5x + 2y) = 2(8) \\ -5(2x + 3y) = -5(12) \end{cases}; x = 0;$$

$$-11y = -44; \quad 5x + 2(4) = 8; y = 4$$

Discussion Guide

To support students in reflecting on the activities and to gather some formative information about student learning, use the following prompts to facilitate a class discussion to "debrief" the station activities.

Prompts/Questions

1. How do you solve a system of linear equations by addition?

2. What is an example of a system of linear equations that you can solve by addition?

3. How do you solve a system of linear equations by multiplication and addition?

4. What is an example of a system of linear equations that you can solve by multiplication and addition?

5. Does it matter which variable you cancel out through multiplication and/or addition? Why or why not?

6. What type of relationship must the same variable in each equation have before you can cancel that variable?

Think, Pair, Share

Have students jot down their own responses to questions, then discuss with a partner (who was not in their station group), and then discuss as a whole class.

Suggested Appropriate Responses

1. Add the same variable in both equations to cancel it out. Solve for the other variable. Then solve for the original variable.

2. Answers will vary. Possible answer: $\begin{cases} x + 2y = 4 \\ x - 2y = 10 \end{cases}$

3. Multiply the same variable in each equation so they have coefficients of opposite values. Then use addition to cancel out that variable. Solve for the other variable. Then solve for the original variable.

4. Answers will vary. Possible answer: $\begin{cases} -2x + y = 2 \\ x + 2y = 10 \end{cases}$

5. No, it doesn't matter. You will arrive at the same answer.

6. The same variable must have opposite coefficients in the two equations.

Possible Misunderstandings/Mistakes

- Not making sure that the same variable has opposite coefficients in both equations before canceling it out

- When using multiplication, forgetting to apply it to all terms in the equation

- Selecting the wrong multipliers to multiply by each equation in order to cancel out a variable

- Trying to solve the system without canceling out a variable through the elimination method

Reasoning with Equations and Inequalities
Set 7: Solving 2-by-2 Systems by Elimination

Station 1

Use what you know about systems of linear equations to answer the questions.

1. How do you solve the equation $x + x + 6 = 14$?

2. What is the solution to this equation? _____

You can apply this same technique when solving systems of linear equations.

Given: $\begin{cases} x + y = 11 \\ x - y = 5 \end{cases}$

You can add these equations together by adding like terms.

3. Which variable would cancel out? Explain your answer.

4. In the space below, rewrite the variable and constant that is left after adding the two equations together. Then solve for that variable.

    _____

5. What is the value of the other variable?

    _____

6. What is the solution to $\begin{cases} x + y = 11 \\ x - y = 5 \end{cases}$?

    _____

continued

Reasoning with Equations and Inequalities
Set 7: Solving 2-by-2 Systems by Elimination

As a group, solve the following systems of linear equations using this addition method. Show your work.

7. $\begin{cases} 2x + y = 10 \\ -2x + 2y = 14 \end{cases}$

8. $\begin{cases} -x - y = 10 \\ x - y = 8 \end{cases}$

Algebra I Station Activities for Common Core State Standards

Reasoning with Equations and Inequalities
Set 7: Solving 2-by-2 Systems by Elimination

Station 2

At this station, you will find 10 yellow, 10 red, and 20 green algebra tiles. You will also be given index cards with the symbols x, y, $-$, $+$, and $=$ written on them.

Use the yellow algebra tiles to represent the coefficient of x. Use the red algebra tiles to represent the coefficient of y. Use the green algebra tiles to represent the constant.

As a group, arrange the algebra tiles and the x, y, $-$, $+$, and $=$ index cards to model the following system of linear equations:

$$\begin{cases} 3x + 2y = 8 \\ 4x - 2y = 6 \end{cases}$$

To solve this system of linear equations you add the two equations.

1. Which algebra tiles will be added together?

    _____

2. Which algebra tiles will cancel each other out?

    _____

3. Arrange the algebra tiles to model the new equation you found. Draw a picture of this equation in the space below.

4. Move the algebra tiles around to solve for the remaining variable. What is this variable and its solution?

    _____

continued

Reasoning with Equations and Inequalities
Set 7: Solving 2-by-2 Systems by Elimination

5. How can you find the value of the other variable?

6. What is the solution to this system of linear equations?

    _____

7. Use your algebra tiles to model and solve the following system of linear equations. Write your answer in the space below.

    $$\begin{cases} -3x + y = 10 \\ 3x + y = 22 \end{cases}$$

Reasoning with Equations and Inequalities
Set 7: Solving 2-by-2 Systems by Elimination

Station 3

Using the elimination method to solve a system of equations means that you eliminate one variable so you can solve for the other variable. Use this information to answer the questions about the following system of equations:

$$\begin{cases} 4x + 10y = 42 \\ 2x + y = 5 \end{cases}$$

1. Can you eliminate a variable from the system of equations above given the form it is written in? Why or why not?

2. What number can you multiply x by in the second equation in order to cancel out the x variable?

   _____

3. Cancel out the x variable and solve for y. Show your work. Remember to keep both sides of the equation balanced.

4. Solve for x. What is the solution to this system of equations? _____

5. Find the solution for the same system $\begin{cases} 4x + 10y = 42 \\ 2x + y = 5 \end{cases}$, but this time eliminate the y variable first. Show your work.

Reasoning with Equations and Inequalities
Set 7: Solving 2-by-2 Systems by Elimination

Station 4

At this station, you will find eight index cards with the following written on them:

$$5x = 0; \quad \begin{cases} 10x + 4y = 16 \\ -10x - 15y = -60 \end{cases}; \quad (0, 4); \quad \begin{cases} 2(5x + 2y) = 2(8) \\ -5(2x + 3y) = -5(12) \end{cases}; \quad x = 0; \; -11y = -44;$$

$$5x + 2(4) = 8; \; y = 4$$

Each index card represents a step in solving a system of linear equations. Shuffle the index cards. Work as a group to arrange the index cards in the correct order to solve the following system of linear equations:

$$\begin{cases} 5x + 2y = 8 \\ 2x + 3y = 12 \end{cases}$$

Write the steps on the lines below.

1. _____

2. _____

3. _____

4. _____

5. _____

6. _____

7. _____

8. _____

9. What strategy did you use to arrange the index cards?

Reasoning with Equations and Inequalities

Set 8: Using Systems in Applications

Goal: To provide opportunities for students to develop concepts and skills related to solving systems of linear equations using graphing, substitution, and elimination methods

Common Core Standards

Algebra: Creating Equations★

Create equations that describe numbers or relationships.

A-CED.2. Create equations in two or more variables to represent relationships between quantities; graph equations on coordinate axes with labels and scales.

A-CED.3. Represent constraints by equations or inequalities, and by systems of equations and/ or inequalities, and interpret solutions as viable or nonviable options in a modeling context.

Algebra: Reasoning with Equations and Inequalities

Solve systems of equations.

A-REI.5. Prove that, given a system of two equations in two variables, replacing one equation with the sum of that equation and a multiple of the other produces a system with the same solutions.

A-REI.6. Solve systems of linear equations exactly and approximately (e.g., with graphs), focusing on pairs of linear equations in two variables.

Represent and solve equations and inequalities graphically.

A-REI.11. Explain why the x-coordinates of the points where the graphs of the equations $y = f(x)$ and $y = g(x)$ intersect are the solutions of the equation $f(x) = g(x)$; find the solutions approximately; e.g., using technology to graph the functions, make tables of values, or find successive approximations. Include cases where $f(x)$ and/or $g(x)$ are linear, polynomial, rational, absolute value, exponential, and logarithmic functions.★

Student Activities Overview and Answer Key

Station 1

Students will be given a real-world application of systems of linear equations. They will set up the system of linear equations. Then they will find the solution using substitution, graphing, or elimination. They will explain the strategy behind the method they used to find the solution.

Answers

1. $x + y = 100$

2. $10x$

3. $25y$

4. $10x + 25y = 2050$

5. $\begin{cases} x + y = 100 \\ 10x + 25y = 2050 \end{cases}$

6. Answers will vary. Possible answer:
$$y = 100 - x$$
$$10x + 25(100 - x) = 2050$$
$$x = 30$$
$$y = 70$$

7. Answers will vary. Possible answer: We used the substitution method because it was easy to solve for y to begin the process.

Station 2

Students will be given six index cards with the following equations written on them:

$$14x + y = 233;\ x + y = 233;\ 14x + 20y = 3748;\ x + 20y = 2128;$$
$$x + y = 2128;\ 14x + 20y = 233$$

Students will work together to solve a real-world application of a system of linear equations. Students will choose which of the two index cards represents the equations in the system. They will solve the system of linear equations using either substitution or elimination. Then they will explain why graphing the system to find the solution is not the best way to solve the problem.

Answers

1. $\begin{cases} x + y = 233 \\ 14x + 20y = 3748 \end{cases}$

2. Number of T-shirts + number of sweatshirts = 233, so $x + y = 233$

Cost per T-shirt = $14, so $14x$

Cost per sweatshirt = $20, so $20y$

Total made on T-shirts and sweatshirts = $3,748, so $14x + 20y = 3748$

3. Answers will vary. Possible answer: Solving by elimination;

$-14(x + y) = -14(233)$

$14x + 20y = 3748$

$y = 81; x = 152$

4. 152 T-shirts

5. 81 sweatshirts

6. Graphing by hand wouldn't be a good choice because the numbers are large.

Station 3

Students will be given a graphing calculator. They will set up a system of linear equations for a real-world example. They will use the graphing calculator to find the solution to the real-world example.

Answers

1. y = number of paperbacks

2. $x + y = 225$

3. $7y$

4. $16x$

5. $16x + 7y = 2583$

6. $\begin{cases} x + y = 225 \\ 16x + 7y = 2583 \end{cases}$

7. I typed $Y_1 = -x + 225$ and $Y_2 = \dfrac{-16}{7}x + 369$. I hit "GRAPH". I hit "2nd", "Trace", and selected "5: Intersect". Then I hit "ENTER" three times. This gave me a solution of $x = 112$ and $y = 113$.

8. 112

9. 113

Station 4

Students will be given a real-world application of systems of linear equations that involves the $d = rt$ formula. They will set up the system of linear equations. Then they will solve the system to find two rates and distances traveled.

Answers

1. Tom's speed

2. $x = y + 15$

3. $2x$

4. $2y$

5. $2x + 2y = 300$

6. $\begin{cases} x = y + 15 \\ 2x + 2y = 300 \end{cases}$

7. Answers will vary. Possible answer: Substitution method;

$$2(y + 15) + 2y = 300$$
$$y = 67.5$$
$$x = 82.5$$

8. 82.5 mph

9. 67.5 mph

10. Tom = 165 miles and Jeremy = 135 miles

Materials List/Setup

Station 1   none

Station 2   six index cards with the following equations written on them:

$14x + y = 233$; $x + y = 233$; $14x + 20y = 3748$; $x + 20y = 2128$;

$x + y = 2128$; $14x + 20y = 233$

Station 3   graphing calculator

Station 4   none

Discussion Guide

To support students in reflecting on the activities and to gather some formative information about student learning, use the following prompts to facilitate a class discussion to "debrief" the station activities.

Prompts/Questions

1. What is the first step you should take when setting up a system of linear equations from a real-world example?

2. How can you find the coefficients of the variables in the equations you use based on a real-world example?

3. To solve for two variables, how many equations do you need in your system of linear equations?

4. Will solving systems of linear equations by substitution, elimination, or graphing give you the same solution? Explain your answer.

5. In what situation would you want to avoid solving a system of linear equations by graphing it by hand?

Think, Pair, Share

Have students jot down their own responses to questions, then discuss with a partner (who was not in their station group), and then discuss as a whole class.

Suggested Appropriate Responses

1. Define what the x and y variables represent.

2. Answers will vary. Possible answer: Look for like units such as $.

3. Two equations; the number of equations matches the number of variables.

4. Yes, they will give you the same solution just reached in a different manner.

5. when you have large numbers

Possible Misunderstandings/Mistakes

- Not defining what each variable represents
- Only using one equation instead of two equations to find the solution
- Incorrectly setting up each equation in the system of linear equations
- Not recognizing that similar units (such as $) should be in the same equation
- Incorrectly using substitution, elimination, and graphing methods to find the solution

Reasoning with Equations and Inequalities
Set 8: Using Systems in Applications

Station 1

At this station, you will find a real-world application of systems of linear equations. Work together to answer the questions.

> A science teacher wants to buy 100 new test tubes for the classroom lab. Acme and Zenith make two different types of test tubes. Acme test tubes cost $10 each. Zenith test tubes cost $25 each. The science teacher has $2,050 to spend. How many of each test tube should she buy?

Let x represent the number of test tubes the science teacher buys from Acme Test tubes. Let y represent the number of test tubes she buys from Zenith.

1. Write an equation that represents the total number of test tubes the science teacher will buy.

   _____

2. Write a term that represents "Acme test tubes cost $10 each." _____

3. Write a term that represents "Zenith test tubes cost $25 each." _____

4. Use the terms you found in problems 2 and 3 to write an equation that represents the amount of money the science teacher has to spend.

   _____

5. Write the two equations you found as a system of linear equations.

continued

Reasoning with Equations and Inequalities
Set 8: Using Systems in Applications

6. Solve the system of linear equations by substitution, graphing, OR elimination. Show your work.

7. Which method did you use to solve the system of linear equations? Why?

Reasoning with Equations and Inequalities
Set 8: Using Systems in Applications

Station 2

At this station, you will find six index cards with the following equations written on them:

$$14x + y = 233; \quad x + y = 233; \quad 14x + 20y = 3748; \quad x + 20y = 2128;$$

$$x + y = 2128; \quad 14x + 20y = 233$$

Shuffle the cards and place them face-down. As a group, read the following real-world problem scenario:

> For a school fundraiser, students sold T-shirts with the school logo for $14. They sold sweatshirts with the school logo for $20. They sold 233 pieces of clothing and raised $3,748. How many of each type of shirt did they sell at the fundraiser?

Let x = the number of T-shirts sold. Let y = the number of sweatshirts sold.

Flip over the index cards. Work together to find the two equations that represent the system of equations for this real-world application.

1. Write the system of linear equations in the space below.

2. Explain why you chose each equation.

continued

Reasoning with Equations and Inequalities
Set 8: Using Systems in Applications

3. Solve the system of linear equations by either substitution or elimination. Show your work in the space below.

4. How many T-shirts were sold? _____

5. How many sweatshirts were sold? _____

6. Explain why solving this system of equations by constructing a graph by hand isn't the best way to solve this problem.

Reasoning with Equations and Inequalities
Set 8: Using Systems in Applications

Station 3

At this station, you will find a graphing calculator. Work as a group to set up a system of linear equations for a real-world problem. Then use the graphing calculator to find the solution to the system of linear equations. You might need to use the Zoom feature or set the viewing window to see the intersection point.

> A bookstore sold 225 books that were either hardcover or paperback. The cost of a paperback book is $7. The cost of a hardcover book is $16. The bookstore made $2,583. How many of each type of book did the bookstore sell?

1. If you let x = the number of hardcover books sold, then what can you let y equal?

    _____

2. Write an equation that represents, "A bookstore sold 225 books that were either hardcover or paperback."

    _____

3. Write a term that represents "The cost of a paperback is $7." _____

4. Write a term that represents "The cost of a hardcover is $16." _____

5. Write an equation that represents the number of hardcovers and paperbacks sold to make $2,583.

    _____

6. Write the system of linear equations you have created.

continued

Reasoning with Equations and Inequalities
Set 8: Using Systems in Applications

7. Use your graphing calculator to find the number of hardcovers and paperbacks sold. Explain how you used your graphing calculator to find this answer.

8. How many hardcovers were sold? _____

9. How many paperbacks were sold? _____

Reasoning with Equations and Inequalities
Set 8: Using Systems in Applications

Station 4

Work together to set up a system of linear equations for the problem scenario below. Then find the solution for the system.

> Tom and Jeremy are 300 miles apart, driving to meet each other at a certain restaurant. Tom drives 15 miles per hour faster than Jeremy. They arrive at the same point after 2 hours. How fast is each person driving?

1. If you let y = Jeremy's speed, what can you let x equal? _____

2. Write x in terms of y. _____

This is the first equation in your system of equations.

3. The distance formula is *distance* = *rate* • *time*. If the time is 2 hours, how you can you express the distance Tom traveled?

   _____

4. How can you express the distance Jeremy traveled? _____

5. Write an equation that represents "The distance Tom traveled + the distance Jeremy traveled is equal to 300 miles."

This is the second equation in your system of linear equations.

6. Write the system of linear equations in the space below.

continued

Reasoning with Equations and Inequalities
Set 8: Using Systems in Applications

7. Find the solution to the system of linear equations using the substitution or elimination method. Show your work in the space below.

8. How fast was Tom driving? _____

9. How fast was Jeremy driving? _____

10. How many miles did each person travel?

Reasoning with Equations and Inequalities

Goal: To provide opportunities for students to develop concepts and skills related to solving systems of linear inequalities, including real-world problems through graphing two and three variables by hand and using a graphing calculator

Common Core Standards

Algebra: Creating Equations★

Create equations that describe numbers or relationships.

A-CED.3. Represent constraints by equations or inequalities, and by systems of equations and/or inequalities, and interpret solutions as viable or nonviable options in a modeling context.

Algebra: Reasoning with Equations and Inequalities

Represent and solve equations and inequalities graphically.

A-REI.12. Graph the solutions to a linear inequality in two variables as a half-plane (excluding the boundary in the case of a strict inequality), and graph the solution set to a system of linear inequalities in two variables as the intersection of the corresponding half-planes.

Student Activities Overview and Answer Key

Station 1

Students will be given graph paper and a ruler. Students will graph two inequalities one at a time on the same graph. Then they will view the two-variable inequalities as a system. They will derive how to find the solutions of a system of linear inequalities. Then they will solve the system of linear inequalities and highlight the solutions with a yellow highlighter.

Answers

1.

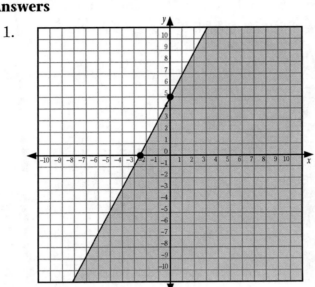

2. $(-5/2, 0)$

3. $(0, 5)$

4. below

5. It should have a solid line because of the ≤ sign.

6.

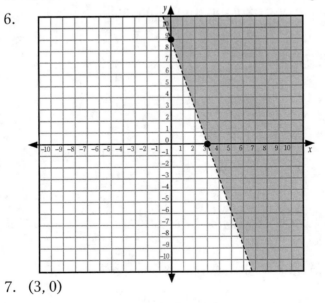

7. $(3, 0)$

8. $(0, 9)$

9. above

10. It should have a dashed line because of the > sign.

Algebra I Station Activities for Common Core State Standards

11. Graph both inequalities on the same graph. Shade solutions for both inequalities. The points where the shaded regions overlap are solutions to the system of linear inequalities.

12.

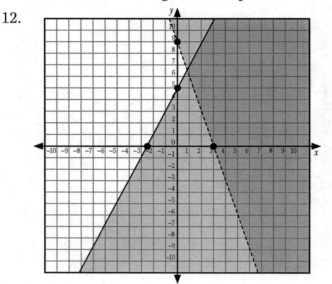

Station 2

Students will be given the graph of a system of linear inequalities in two variables. They will derive each inequality. Then they will use a highlighter to depict the solutions to the system of linear inequalities. They will also find the solutions of the system of three linear inequalities in two variables.

Answers

1. x-intercept $(2, 0)$; y-intercept $(0, -10)$; $m = 5$

2. $y > 5x - 10$

3. x-intercept $(-6, 0)$; y-intercept $(0, 12)$; $m = 2$

4. $y < 2x + 12$

5. $\begin{cases} y > 5x - 10 \\ y < 2x + 12 \end{cases}$

6. The region where the two shaded regions overlap contains the solutions to the system of linear inequalities.

7. Make sure they highlight the region where the two shaded regions overlap.

8. Make sure they highlight the region where the two shaded regions overlap.

9. Make sure they highlight the region where all three shaded regions overlap. This region will look like a triangle.

10. In problem 9, you have three inequalities in the system instead of just two.

Station 3

Students will be given a graphing calculator and yellow highlighter. Students will use the graphing calculator to analyze, graph, and find the solutions of a system of linear inequalities in two variables. Then they use the graphing calculator to graph and find the solutions of a system of three linear inequalities in two variables.

Answers

1. Use ◣ because of the ≤ sign.

2. Use ◥ because of the > sign.

3. the region where the two shaded regions overlap

4.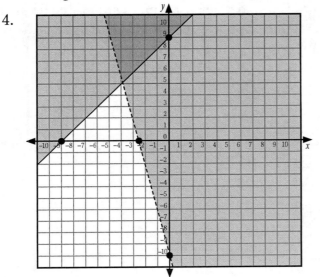

5.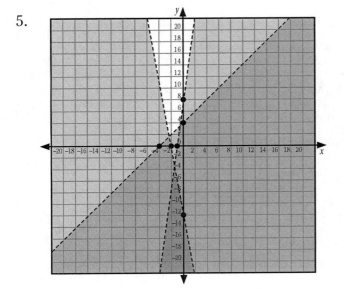

Station 4

Students will be given a real-world application of a system of linear inequalities. Students will derive a system of linear inequalities, graph it, and then find the region of possible solutions. Then they will use this region in a linear equation to find maximum profit.

Answers

1. y = number of plastic doghouses produced in one week

2. $5x + 2y \leq 50$; (time it takes to produce one wood doghouse • number of wood doghouses produced each week) + (time it takes to produce one plastic doghouse • number of plastic doghouses produced each week) ≤ maximum amount of production time for building each week

3. $x + 2y \leq 30$; (time it takes to sand and paint one wood doghouse • number of wood doghouses produced each week) + (time it takes to assemble one plastic doghouse • number of plastic doghouses produced each week) ≤ maximum amount of production time for sanding, painting, and assembling

4. $\begin{cases} 5x + 2y \leq 50 \\ x + 2y \leq 30 \end{cases}$

5.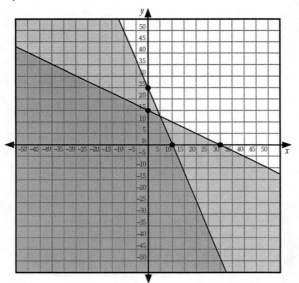

6. You can't build a negative number of doghouses.

7. (0, 0), (0, 15), (10, 0), and (7, 8)

8. $S = 40x + 20y$

9.

x	y	Sales equation	Sales ($)
0	0	$S = 40(0) + 20(0) = 0$	$0
0	15	$S = 40(0) + 20(15) = 300$	$300
10	0	$S = 40(10) + 20(0) = 400$	$400
7	8	$S = 40(7) + 20(8) = 440$	$440

10. (7, 8)

11. The company should produce 7 wood doghouses and 8 plastic doghouses each week to maximize their sales at $440.

Materials List/Setup

Station 1   graph paper; ruler; yellow highlighter

Station 2   yellow highlighter

Station 3   graphing calculator; yellow highlighter

Station 4   graph paper; ruler

Discussion Guide

To support students in reflecting on the activities and to gather some formative information about student learning, use the following prompts to facilitate a class discussion to "debrief" the station activities.

Prompts/Questions

1. How do you solve a system of linear inequalities with two variables?

2. Why does a system of linear inequalities have more than one solution?

3. How do you solve a system of three linear inequalities with two variables?

4. How do you create a system of linear inequalities from a real-world application?

5. Give an example of when you can use both a system of linear inequalities and an equation to solve a real-world application.

6. Why in most real-world applications of linear inequalities must the variables be nonnegative?

Think, Pair, Share

Have students jot down their own responses to questions, then discuss with a partner (who was not in their station group), and then discuss as a whole class.

Suggested Appropriate Responses

1. Graph both inequalities and shade the region that contains their solutions. The solution of the system of linear inequalities is the region where the two solution regions of the inequalities overlap.

2. An inequality means that there is more than one answer that could satisfy the equation.

3. Graph all three inequalities and shade the regions that contain their solutions. The solution of the system of linear inequalities is the region where the solution regions of the inequalities overlap.

4. Identify the variables. Write the inequalities based on the given constraints.

5. Sample answer: When you have a system of linear inequalities that represents a production schedule. The accompanying equation can represent money made from the production schedule.

6. You can't create negative amounts of products, time, or distance.

Possible Misunderstandings/Mistakes

- Not identifying variables used in inequalities
- Incorrectly graphing each inequality by shading the incorrect region of solutions
- Not realizing that the solution to the system of the linear inequalities is the area where the shaded regions overlap
- Only graphing two inequalities in a two variable system of three linear inequalities
- Not realizing that you can't have negative amounts of real-world amounts, such as products, time, or distance

Reasoning with Equations and Inequalities
Set 9: Solving Systems of Inequalities

Station 1

At this station, you will find graph paper, a highlighter, and a ruler. Work together to graph each inequality.

1. On your graph paper, graph $y \leq 2x + 5$.

2. What is the x-intercept of the graph? _____

3. What is the y-intercept of the graph? _____

4. Shade the graph to indicate the possible solutions to the inequality.

   Are the solutions to the inequality above or below the line? _____

5. Should your line be dashed or solid? Explain your answer.

6. Graph $y > -3x + 9$ on the same graph you used in problem 1.

7. What is the x-intercept of the graph? _____

8. What is the y-intercept of the graph? _____

9. Shade the graph to indicate the possible solutions to the inequality.

   Are the solutions to the inequality above or below the line? _____

continued

Reasoning with Equations and Inequalities
Set 9: Solving Systems of Inequalities

10. Should your line be dashed or solid? Explain your answer.

11. By graphing the two inequalities on the same graph, you have created a system of linear inequalities. How can you solve this system of linear inequalities?

12. Use your yellow highlighter to show the solutions to $\begin{cases} y \leq 2x + 5 \\ y > -3x + 9 \end{cases}$.

© 2011 Walch Education

Reasoning with Equations and Inequalities
Set 9: Solving Systems of Inequalities

Station 2

At this station, you will find a yellow highlighter and a graph of a system of linear inequalities. Work together to analyze the graph and find the solutions to the system of linear inequalities.

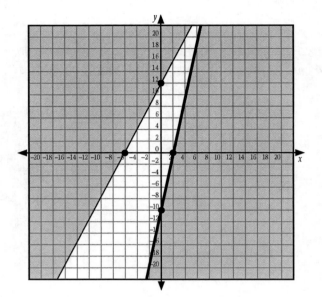

1. What are the *x*-intercept, *y*-intercept, and slope of the HEAVY line?

   x-intercept: _____

   y-intercept: _____

   Slope: _____

2. Use the information in problem 1 to find the linear inequality for the HEAVY line. Show your work in the space below.

3. What are the *x*-intercept, *y*-intercept, and slope of the LIGHT line?

   x-intercept: _____

   y-intercept: _____

   Slope: _____

continued

4. Use the information in problem 3 to find the linear inequality for the LIGHT line. Show your work in the space below.

5. In the space below, write the inequalities as a system of linear inequalities.

6. How can you use the graph to find the solutions of this system of linear inequalities?

7. Use your highlighter to show the solutions to the system of linear inequalities.

For problems 8 and 9, use your highlighter to solve the system of linear inequalities in each graph.

8.

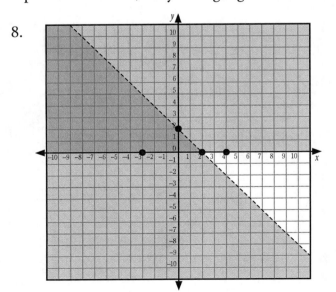

9.

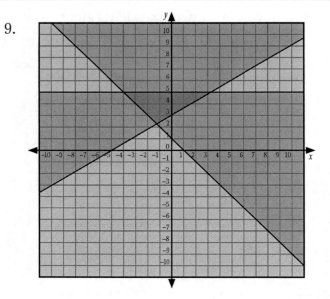

10. Why does the region containing the solutions appear different in problem 9 than in problem 8?

_____

Reasoning with Equations and Inequalities
Set 9: Solving Systems of Inequalities

Station 3

At this station, you will find a graphing calculator and a yellow highlighter. Work as a group to follow these steps in order to graph a system of linear inequalities on your graphing calculator.

$$\text{Given: } \begin{cases} y \leq x + 9 \\ y > -4x - 10 \end{cases}$$

Steps to graph the system of inequalities on a graphing calculator:

Step 1: Hit the "Y=" key. Set Y1 = x + 9. Use the left arrow to go to the far left of Y1. Hit "ENTER" until a blinking ◣ appears.

1. Why do you think you use the ◣ icon next to Y1?

Step 2: Set Y2 = −4x − 10. Use the left arrow to go to the far left of Y2. Hit "ENTER" until a blinking ◤ appears.

2. Why do you think you use the ◤ icon next to Y2?

Step 3: Hit the "GRAPH" key.

3. Look at the graph of the system of inequalities. Which region contains the solutions to the system of linear inequalities? Explain your answer.

continued

Reasoning with Equations and Inequalities
Set 9: Solving Systems of Inequalities

4. Sketch the graph of the system of inequalities in the space below. Use the yellow highlighter to show the region that contains the solutions to the system.

Use your graphing calculator to find the solutions to the following system of linear inequalities:

$$\begin{cases} y \leq x + 4 \\ y > 6x - 12 \\ y < 2x + 8 \end{cases}$$

5. In the space below, sketch the system of inequalities. Use your yellow highlighter to show the solutions to the system of linear inequalities.

Reasoning with Equations and Inequalities
Set 9: Solving Systems of Inequalities

Station 4

At this station, you will find graph paper and a ruler. Work as a group to graph and solve a real-world application of a system of linear inequalities.

The Profitable World of Doghouses

A pet products company sells wood doghouses and plastic doghouses. It takes the company 5 hours to build a wood doghouse and 2 hours to build a plastic doghouse. The company builds these doghouses during a maximum production time of 50 hours per week.

It takes an additional 1 hour to sand and paint each wood doghouse. It takes an additional 2 hours to assemble each plastic doghouse. The company sands, paints, and assembles these doghouses during a maximum production time of 30 hours per week.

The wood doghouses sell for $40. The plastic doghouses sell for $20.

How many wood doghouses and plastic doghouses should the company produce each week in order to make the most money?

1. If x = the number of wood doghouses produced each week, then what is y?

   _____

2. Write an inequality that represents the production time needed to build each doghouse. Explain your answer.

3. Write an inequality that represents the production time needed for sanding, painting, and assembling each doghouse. Explain your answer.

4. In the space below, write the system of linear inequalities you have created.

continued

Reasoning with Equations and Inequalities
Set 9: Solving Systems of Inequalities

5. Use your graph paper and ruler to graph this system of linear inequalities.

6. Why must you only look at solutions where $x > 0$ and $y > 0$?

7. Identify four points that bound the region of possible solutions to the system of linear inequalities. (*Hint*: For the fourth point, approximate the intersection of the two lines.)

8. Write an equation that will represent the amount of money made for each doghouse. Explain your answer.

9. Use your answer in problem 8 as the "Sales equation."

   Complete the table below.

x	y	Sales equation	Sales ($)

10. Which point yields the highest profit?

    _____

11. Based on your answer to problem 10, how many wood doghouses and plastic doghouses should the company produce in order to maximize sales? Explain your answer.

Reasoning with Equations and Inequalities

Goal: To provide opportunities for students to develop concepts and skills related to solving quadratic equations using the square root property. The quadratic equations will have two solutions, one solution, or no real solutions.

Common Core Standards

Algebra: Reasoning with Equations and Inequalities

Solve equations and inequalities in one variable.

A-REI.4. Solve quadratic equations in one variable.

b. Solve quadratic equations by inspection (e.g., for $x^2 = 49$), taking square roots, completing the square, the quadratic formula and factoring, as appropriate to the initial form of the equation. Recognize when the quadratic formula gives complex solutions and write them as $a \pm bi$ for real numbers a and b.

Student Activities Overview and Answer Key

Station 1

Students will be given a number cube. Students will use the number cube to populate quadratic equations that can be solved by using the square root property. They will realize that quadratic equations solved by the square root property have answers which are the +/− of the same number.

Answers

1. Answers will vary. Make sure they add the constant to both sides of the equation.

2. Answers will vary. Make sure they take square root of both sides of the equation.

3. Answers will vary. Make sure they find the +/− square root of the constant.

4. Two, because the x^2 exponent tells you that there are two solutions.

5. Answers will vary. Make sure that students first add the constant to both sides of the equation, then take the square root of both sides, then add the constant from the left side to the right side of the equation.

6. Answers will vary. Make sure that students first add the constant to both sides of the equation, then take the square root of both sides, then add the constant from the left side to the right side of the equation.

7. Two solutions, because the square root of a number is the +/− of the number; the square of a positive is a positive and the square of a negative is also a positive.

Station 2

Students will be given 12 index cards with the following written on them:

$$a = 1; a = 2; a = 4; a = 5; b = 0; b = 0; b = 0; b = 0; c = 16; c = 25; c = 36; c = 49$$

Students work together to set up quadratic equations and solve them using the square root property.

Answers

1. Answers will vary. Possible answer: $a = 1$, $b = 0$, and $c = 16$, so

   $$x^2 - 16 = 0$$
   $$x^2 = 16$$
   $$x = \pm 4$$

2. Answers will vary. Possible answer: $a = 2$, $b = 0$, and $c = 36$, so

   $$2x^2 - 36 = 0$$
   $$2x^2 = 36$$
   $$x^2 = 18$$
   $$x = \pm 3\sqrt{2}$$

3. Answers will vary. Possible answer: $a = 5$, $b = 0$, and $c = 25$, so

   $$5x^2 = 25$$
   $$x^2 = 5$$
   $$x = \pm\sqrt{5}$$

4. Answers will vary. Possible answer: $a = 4$, $b = 0$, and $c = 49$, so

   $$4x^2 = 49$$
   $$x^2 = \frac{49}{4}$$
   $$x = \pm\sqrt{\frac{49}{4}} = \pm\frac{7}{2}$$

5. two solutions, because you find the square root

Station 3

Students will be given a deck of playing cards that contain only the numbers 2–10. Students will draw numbers and use them to populate quadratic equations. They will solve the quadratic equations using the square root property. They will realize that the square root of a negative number yields no real solutions. They will also realize that the square root of 0 yields only one solution.

Answers

1. Answers will vary. Make sure students subtract the constant from both sides of the equation.

2. negative, because of subtraction

3. You can't find the square root of a negative number in the real number system.

4. no real solutions

5. Answers will vary. Make sure students subtract the constant from both sides of the equation.

6. It cancels out, so the right-hand side of the equation is 0.

7. $x = \pm\sqrt{0} = 0$

8. $\pm 0 = 0$

9. $3x^2 - 6 = -6$

   $3x^2 = 0$

   $x = 0$

10. $5x^2 + 125 = 0$

    $5x^2 = -125$

    $x^2 = -25$

    $x = $ no real solutions

Station 4

Students will be given five index cards with the following written on them:

$$2x^2 = 72\,;\ 5x^2 - 2x - 32 = 3x^2 - 2x + 40\,;\ x^2 = 36\,;\ x = \pm 6\,;$$
$$5x^2 - 3x^2 - 2x + 2x = 32 + 40$$

Students will arrange the index cards in order to demonstrate how to find the solution to a quadratic equation that is not given in standard form. Students will indicate their strategy and acknowledge the square root property. Then they will solve quadratic equations not given in standard form using the square root property.

Answers

1. $5x^2 - 2x - 32 = 3x^2 - 2x + 40$

2. $5x^2 - 3x^2 - 2x + 2x = 32 + 40$

3. $2x^2 = 72$

4. $x^2 = 36$

5. $x = \pm 6$

6. square root property

7. Answers will vary.

8. $2x^2 + 3x = 144 + 3x + x^2$

    $x^2 = 144$

    $x = \pm 12$

9. $10x^2 - 4x + 16 + 4x = 16$

    $10x^2 = 0$

    $x = 0$

10. $x^2 + 3x = -x^2 + 3x - 8$

    $2x^2 = -8$

    $x^2 = -4$

    $x = $ no real solutions

Materials List/Setup

Station 1   number cube

Station 2   12 index cards with the following written on them:

   $a = 1$; $a = 2$; $a = 4$; $a = 5$; $b = 0$; $b = 0$; $b = 0$; $b = 0$; $c = 16$; $c = 25$; $c = 36$; $c = 49$

Station 3   deck of playing cards that contain only the numbers 2–10

Station 4   five index cards with the following written on them:

   $2x^2 = 72$; $5x^2 - 2x - 32 = 3x^2 - 2x + 40$; $x^2 = 36$; $x = \pm 6$;

   $5x^2 - 3x^2 - 2x + 2x = 32 + 40$

Discussion Guide

To support students in reflecting on the activities and to gather some formative information about student learning, use the following prompts to facilitate a class discussion to "debrief" the station activities.

Prompts/Questions

1. How many solutions does a quadratic equation have?

2. If the standard form of a quadratic equation is $ax^2 + bx + c = 0$, which letter (a, b, or c) should be equal to 0 in order to use the square root property?

3. If $x^2 = $ a negative number, how many solutions does the equation have in the real number system?

4. If $x^2 = 0$, how many solutions does the equation have?

5. Why does the square root property yield answers that are the ± version of the same number?

Think, Pair, Share

Have students jot down their own responses to questions, then discuss with a partner (who was not in their station group), and then discuss as a whole class.

Suggested Appropriate Responses

1. two solutions

2. $b = 0$

3. no real solutions

4. one solution, which equals zero

5. Because there are two possible solutions, a positive number can be squared to equal a positive number, and a negative number can be squared to yield that same positive number. For example, 3 squared is 9, and so is (−3) squared.

Possible Misunderstandings/Mistakes

- Not realizing that there are two solutions when $x^2 = $ a positive number
- Not realizing that there are no real solutions when $x^2 = $ a negative number
- Not realizing that there is only one solution when $x^2 = 0$
- Incorrectly combining like terms before solving an equation

Reasoning with Equations and Inequalities
Set 10: Solving Quadratic Equations by Finding Square Roots

Station 1

At this station, you will find a number cube. Roll the number cube and write the number in the first box below. Roll it again, and write the second number in the second box.

$$x^2 - \boxed{}\,\boxed{} = 0$$

As a group, solve this quadratic equation by finding the value of x.

1. What is the first step you take to solve this equation?

2. What is the second step you take to solve this equation?

3. What is the solution to this equation? _____

4. How many solutions did you find? Explain your answer.

Given: $(x - \boxed{})^2 - \boxed{}\,\boxed{} = 0$

Roll the number cube and write the number in the box above. Repeat this process until each box contains a number.

5. Solve this equation. Show your work.

continued

Reasoning with Equations and Inequalities
Set 10: Solving Quadratic Equations by Finding Square Roots

6. What strategy did you use to solve this equation?

7. How many solutions did you find? Explain your answer.

Reasoning with Equations and Inequalities
Set 10: Solving Quadratic Equations by Finding Square Roots

Station 2

At this station, you will find 12 index cards with the following written on them:

$$a = 1; a = 2; a = 4; a = 5; b = 0; b = 0; b = 0; b = 0; c = 16; c = 25; c = 36; c = 49$$

Place the index cards in three piles, organized by their letter: a, b, and c. For each problem, work as a group by choosing a card from each pile and writing it in the form shown. For example, if you draw the cards "$a = 4$," "$b = 0$," and "$c = 16$," inserting them into the form $ax^2 + bx - c$ results in this equation:

$$4x^2 + 0x = 16$$

Then solve each equation for x using the square root property. Show your work.

1. $ax^2 + bx - c = 0$

2. $ax^2 + bx - c = 0$

continued

Reasoning with Equations and Inequalities
Set 10: Solving Quadratic Equations by Finding Square Roots

3. $ax^2 + bx = c$

4. $ax^2 + bx = c$

5. How many solutions did you find for each problem? Explain your answer.

Reasoning with Equations and Inequalities
Set 10: Solving Quadratic Equations by Finding Square Roots

Station 3

At this station, you will find a deck of playing cards that contain only the numbers 2–10. Place the cards in a pile. As a group, select one card. Write the number on the face of the card in the box below.

$$x^2 + \boxed{} = 0$$

1. What is the first step you take to solve for x?

2. Is the constant now positive or negative? Explain your answer.

3. How does the sign of the constant affect finding the square root of x?

4. What is the solution for x? _____

Select another card from the pile. Write the number on the face of the card in each box below.

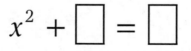

$$x^2 + \boxed{} = \boxed{}$$

5. What is the first step you take to solve for x?

6. What happens to the constant?

continued

Algebra I Station Activities for Common Core State Standards

Reasoning with Equations and Inequalities
Set 10: Solving Quadratic Equations by Finding Square Roots

7. Based on your new constant, what is the value of x? _____

8. Why does x only have one value?

Work together to solve problems 9 and 10. Show your work.

9. $3x^2 - 6 = -6$

10. $5x^2 + 125 = 0$

Reasoning with Equations and Inequalities
Set 10: Solving Quadratic Equations by Finding Square Roots

Station 4

At this station, you will find five index cards with the following written on them:

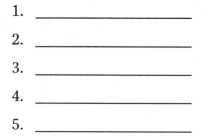

$$2x^2 = 72;\ 5x^2 - 2x - 32 = 3x^2 + 2x + 40;\ x^2 = 36;\ x = \pm 6;$$
$$5x^2 - 3x^2 - 2x + 2x = 32 + 40$$

Shuffle the cards. As a group, arrange the cards to show how to find the solution to $5x^2 - 2x - 32 = 3x^2 - 2x + 40$.

Write the steps on the lines below.

1. _____

2. _____

3. _____

4. _____

5. _____

6. What property did you use to find the solutions for x?

7. What strategy did you use to arrange the cards?

As a group, solve problems 8–10. Show your work.

8. $2x^2 + 3x = 144 + 3x + x^2$

continued

Reasoning with Equations and Inequalities
Set 10: Solving Quadratic Equations by Finding Square Roots

9. $10x^2 - 4x + 16 + 4x = 16$

10. $x^2 + 3x = -x^2 + 3x - 8$

Reasoning with Equations and Inequalities

Set 11: Solving Quadratic Equations Using the Quadratic Formula

Goal: To provide opportunities for students to develop concepts and skills related to solving quadratic equations by factoring and using the quadratic formula

Common Core Standards

Algebra: Reasoning with Equations and Inequalities

Solve equations and inequalities in one variable.

A-REI.4. Solve quadratic equations in one variable.

b. Solve quadratic equations by inspection (e.g., for $x^2 = 49$), taking square roots, completing the square, the quadratic formula and factoring, as appropriate to the initial form of the equation. Recognize when the quadratic formula gives complex solutions and write them as $a \pm bi$ for real numbers a and b.

Student Activities Overview and Answer Key

Station 1

Students will be given a calculator and eight index cards with the following written on them:

$$x^2 + 9x + 20 = 0; \ 2x^2 + 2x - 12 = 0; \ x^2 + x - 2 = 0; \ 6x^2 + 17x - 14 = 0;$$

$$x = \frac{-1 \pm \sqrt{1^2 - 4(1)(-2)}}{2(1)}; \ x = \frac{-17 \pm \sqrt{17^2 - 4(6)(-14)}}{2(6)};$$

$$x = \frac{-2 \pm \sqrt{2^2 - 4(2)(-12)}}{2(2)}; \ x = \frac{-9 \pm \sqrt{9^2 - 4(1)(20)}}{2(1)}$$

Students will work together to match each quadratic equation with the appropriate quadratic formula. Then they will find the solutions to each quadratic equation. They will also explain when using the quadratic formula would be a better way to solve a quadratic equation than by factoring.

Answers

1. $a = 1, b = 9, c = 20$

2. $x^2 + 9x + 20 = 0$

$x = \dfrac{-9 \pm \sqrt{9^2 - 4(1)(20)}}{2(1)}$ and $x = -4, -5$

3. $x^2 + x - 2$

$x = \dfrac{-1 \pm \sqrt{1^2 - 4(1)(-2)}}{2(1)}$ and $x = -2, 1$

4. $2x^2 + 2x - 12$

$x = \dfrac{-2 \pm \sqrt{2^2 - 4(2)(-12)}}{2(2)}$ and $x = 2, -3$

5. $6x^2 + 17x - 14$

$x = \dfrac{-17 \pm \sqrt{17^2 - 4(6)(-14)}}{2(6)}$ and $x = -7/2, 2/3$

6. Use the quadratic formula when factoring is too complicated.

Station 2

Students will be given a number cube. Students will roll the number cube to populate a quadratic equation. They will find the discriminant of the quadratic equation. They will discover how the value of the discriminant determines the number and types of unique solutions for a quadratic equation.

Answers

1. No, sometimes they have a double root or complex conjugate roots.

2. Answers will vary.

3. Answers will vary.

4. $d = 0; d = 0$

5. $d = 4; d > 0$

6. $d = -287; d < 0$

7. There is a double root; there are two different roots; there are two complex conjugate roots.

8. Answers will vary.

Station 3

Students will be given a calculator and real-world application of the quadratic equation. Students will find the discriminant in order to determine the number and types of solutions to the equation. Then they will find the solutions using the quadratic formula. They will explain why only one solution is appropriate in this real-world application.

Answers

1. $a = -6$, $b = 24$, and $c = 12$

2. 864; $d > 0$

3. $d > 0$, so there are two different real roots.

4. $x = \dfrac{-24 \pm \sqrt{24^2 - 4(-6)(12)}}{2(-6)}$

   $x = -0.45, 4.45$

5. 4.45 seconds

6. You can't use the solution of -0.45 seconds because you can't have negative time.

Station 4

Students will be given 10 blue algebra tiles, 15 red algebra tiles, and 20 green algebra tiles. Students will use the algebra tiles to model quadratic equations. Then they will move the algebra tiles to show the equation in standard form. They will try to factor the equation using algebra tiles. Then they will use the quadratic formula to solve the quadratic equation.

Answers

1. 10 red tiles moved to the left to make $10x$; 5 green tiles moved to the left to make 5;

   $2x^2 + 10x + 5 = 0$

2. no

3. $x = \dfrac{-10 \pm \sqrt{10^2 - 4(2)(5)}}{2(2)} = \dfrac{-5 \pm \sqrt{15}}{2}$

4. 1 blue tile moved to the left to make $5x^2$. 15 red tiles moved to the left to make $12x$. 2 green tiles moved to the left to make 2; $5x^2 + 12x + 2 = 0$

5. no

6. $x = \dfrac{-12 \pm \sqrt{12^2 - 4(5)(2)}}{2(5)} = \dfrac{-12 \pm \sqrt{104}}{10} = \dfrac{-12 \pm 2\sqrt{26}}{10} = \dfrac{-6 \pm \sqrt{26}}{5}$

Materials List/Setup

Station 1   calculator and eight index cards with the following written on them:

$$x^2 + 9x + 20 = 0;\ 2x^2 + 2x - 12 = 0;\ x^2 + x - 2 = 0;\ 6x^2 + 17x - 14 = 0;$$

$$x = \frac{-1 \pm \sqrt{1^2 - 4(1)(-2)}}{2(1)};\ x = \frac{-17 \pm \sqrt{17^2 - 4(6)(-14)}}{2(6)};$$

$$x = \frac{-2 \pm \sqrt{2^2 - 4(2)(-12)}}{2(2)};\ x = \frac{-9 \pm \sqrt{9^2 - 4(1)(20)}}{2(1)}$$

Station 2   number cube

Station 3   calculator

Station 4   10 blue algebra tiles; 15 red algebra tiles; 20 green algebra tiles

Discussion Guide

To support students in reflecting on the activities and to gather some formative information about student learning, use the following prompts to facilitate a class discussion to "debrief" the station activities.

Prompts/Questions

1. Why do you have to put a quadratic equation into standard form before you use the quadratic formula?

2. What does the discriminant of a quadratic equation tell you about the type and number of solutions?

3. How do you find the discriminant of a quadratic equation?

4. What is the quadratic formula?

5. When would you use the quadratic formula instead of solving by factoring?

6. What is a real-world application of a quadratic equation?

Think, Pair, Share

Have students jot down their own responses to questions, then discuss with a partner (who was not in their station group), and then discuss as a whole class.

Suggested Appropriate Responses

1. You need to simplify it and put it in standard form so you can use the appropriate values for a, b, and c in the quadratic formula.

2. If the discriminant $= 0$, then the solution is a real double root. If the discriminant > 0, then there are two different real roots. If the discriminant < 0, then there are two complex conjugate roots.

3. Discriminant $= b^2 - 4ac$

4. $x = \dfrac{-b \pm \sqrt{b^2 - 4ac}}{2a}$

5. When the factoring is too complicated and/or the solutions are fractions or decimals.

6. Answers will vary. Sample answer: finding the height or time in the trajectory of a ball

Possible Misunderstandings/Mistakes

- Not writing the equation in standard form before using the quadratic formula
- Using the wrong values for a, b, and c in the quadratic formula
- Using the wrong values for a, b, and c in the discriminant
- Not keeping track of positive and negative signs of a, b, and c in the quadratic formula

Reasoning with Equations and Inequalities
Set 11: Solving Quadratic Equations Using the Quadratic Formula

Station 1

At this station, you will find a calculator and eight index cards with the following written on them:

$$x^2 + 9x + 20 = 0\,;\ 2x^2 + 2x - 12 = 0\,;\ x^2 + x - 2 = 0\,;\ 6x^2 + 17x - 14 = 0\,;$$

$$x = \frac{-1 \pm \sqrt{1^2 - 4(1)(-2)}}{2(1)}\,;\ x = \frac{-17 \pm \sqrt{17^2 - 4(6)(-14)}}{2(6)}\,;$$

$$x = \frac{-2 \pm \sqrt{2^2 - 4(2)(-12)}}{2(2)}\,;\ x = \frac{-9 \pm \sqrt{9^2 - 4(1)(20)}}{2(1)}$$

Shuffle the cards and place them in a pile.

You can solve quadratic equations using the quadratic formula:

$$x = \frac{-b \pm \sqrt{b^2 - 4ac}}{2a}$$

1. Given $x^2 + 9x + 20 = 0$, what are the values of a, b, and c?

   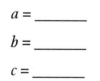

As a group, flip over all the index cards. Work together to match each quadratic equation with the appropriate quadratic formula.

Write each match on the lines that follow. Then solve for x.

2. _____

   $x = $ _____

continued

Reasoning with Equations and Inequalities
Set 11: Solving Quadratic Equations Using the Quadratic Formula

3. _____

   $x =$ _____

4. _____

   $x =$ _____

5. _____

   $x =$ _____

6. When would you use the quadratic formula instead of factoring the quadratic equation? Explain your reasoning.

Reasoning with Equations and Inequalities
Set 11: Solving Quadratic Equations Using the Quadratic Formula

Station 2

At this station, you will find a number cube. As a group, roll the number cube. Write the result in the first box below. Roll the number cube again, and write the new number in the second box.

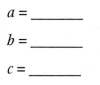

$$x^2 + \boxed{}x + \boxed{} = 0$$

1. Do quadratic equations always have two unique real solutions? Why or why not?

You can determine the number and type of solutions a quadratic equation has by finding its discriminant, D. The discriminant is $b^2 - 4ac$.

2. What are the values of a, b, and c in your quadratic equation above?

   $a =$ _____

   $b =$ _____

   $c =$ _____

3. Find the discriminant for your quadratic equation. Show your work.

4. Find the discriminant for $x^2 + 8x + 16 = 0$. Show your work.

   Is the discriminant > 0, < 0, or 0? _____

   The solution to $x^2 + 8x + 16 = 0$ is $x = -4$, which is a double root.

continued

Reasoning with Equations and Inequalities
Set 11: Solving Quadratic Equations Using the Quadratic Formula

5. Find the discriminant for $x^2 + 3x - 10 = 0$. Show your work.

Is the discriminant > 0, < 0, or 0? _____

The solution to $x^2 + 3x - 10 = 0$ is $x = -5$, $x = 2$, which means the equation has two different real roots.

6. Find the discriminant for $12x^2 - x + 6 = 0$. Show your work.

Is the discriminant > 0, < 0, or 0? _____

The solution to $12x^2 - x + 6 = 0$ has two complex conjugate roots.

7. Based on your answers in problems 4–6, complete these statements:

When the discriminant = 0, _____

_____

When the discriminant > 0, _____

_____

When the discriminant < 0, _____

_____

8. Based on the discriminant you found in problem 3, what type of solution(s) does the quadratic equation you created in problem 1 have?

Reasoning with Equations and Inequalities
Set 11: Solving Quadratic Equations Using the Quadratic Formula

Station 3

At this station, you will find a calculator. The problem scenario below is a real-world application of the quadratic formula. Work as a group to answer the questions.

> At a baseball game, Craig threw a baseball up in the air from a bleacher that was 12 feet off the ground. The initial velocity of the baseball was 24 feet per second. How long will it take for the baseball to hit the ground?

> The height at time t is modeled by $h = -6t^2 + 24t + 12$.

1. What are the values of a, b, and c in the quadratic equation?

   $a =$ _____

   $b =$ _____

   $c =$ _____

2. The discriminant of a quadratic equation is $b^2 - 4ac$. What is the discriminant for the equation above? Show your work.

3. Based on the discriminant you found in problem 2, how many solutions and what type of solution(s) will the quadratic equation have? (*Hint*: When the discriminant = 0, there is one root. When the discriminant > 0, there are two real roots. When the discriminant < 0, there are no real roots.)

continued

Reasoning with Equations and Inequalities
Set 11: Solving Quadratic Equations Using the Quadratic Formula

4. The quadratic formula is $x = \dfrac{-b \pm \sqrt{b^2 - 4ac}}{2a}$. Use this formula to find the solution(s) to $h = -6t^2 + 24t + 12$. Show your work.

5. Which solution gives the amount of time it takes for the baseball to hit the ground?

   _____

6. Why will only one of your solutions equal the time it takes for the baseball to hit the ground?

Reasoning with Equations and Inequalities
Set 11: Solving Quadratic Equations Using the Quadratic Formula

Station 4

You will be given 10 blue algebra tiles, 15 red algebra tiles, and 20 green algebra tiles.

As a group, arrange the algebra tiles to model the following quadratic equation:

$$2x^2 = -10x - 5$$

- Use blue tiles for the coefficient of x^2.
- Use red tiles for the coefficient of x.
- Use green tiles for the constant.

Now, move the algebra tiles to write the equation in standard form.

1. Which algebra tiles did you move and how did you move them?

   Write the equation in standard form: _____

2. Can you arrange the green algebra tiles in the factors below to solve the quadratic equation? Why or why not?

   $(2x + \text{green tiles})(x + \text{green tiles})$

3. Instead of factoring, you can use the quadratic formula:

   $$x = \frac{-b \pm \sqrt{b^2 - 4ac}}{2a}$$

continued

Reasoning with Equations and Inequalities
Set 11: Solving Quadratic Equations Using the Quadratic Formula

Use the quadratic formula to solve the quadratic equation you found in problem 1. Show your work in the space below.

As a group, arrange the algebra tiles to model the following quadratic equation:

$$4x^2 - 3x = -15x^2 - x^2 - 2$$

Move the algebra tiles to write the equation in standard form.

4. Which algebra tiles did you move and how did you move them?

Write the equation in standard form: _____

5. Can you arrange the green algebra tiles in the factors below to solve the quadratic equation? Why or why not?

$(5x + \text{green tiles})(x + \text{green tiles})$

6. Instead of factoring, you can use the quadratic formula:

$$x = \frac{-b \pm \sqrt{b^2 - 4ac}}{2a}$$

Use the quadratic formula to solve the quadratic equation you found in problem 4. Show your work.

Arithmetic with Polynomials and Rational Expressions

Set 1: Operations with Polynomials

Goal: To provide opportunities for students to develop concepts and skills related to adding, subtracting, multiplying, and dividing polynomials

Common Core Standards

Algebra: Arithmetic with Polynomials and Rational Expressions

Perform arithmetic operations on polynomials.

A-APR.1. Understand that polynomials form a system analogous to the integers, namely, they are closed under the operations of addition, subtraction, and multiplication; add, subtract, and multiply polynomials.

Student Activities Overview and Answer Key

Station 1

Students will be given 20 blue algebra tiles, 20 red algebra tiles, 20 green algebra tiles, and 20 yellow algebra tiles. Students work together to model polynomials with algebra tiles. Then they add the polynomials using the algebra tiles.

Answers

1. $8x^2 + xy + 5y^2$

2. Answers will vary. Possible answer: We combined same-color algebra tiles.

3. Answers will vary. Possible answer: We used the zero property to find pairs of the same colored algebra tiles that canceled each other out.

4. zero property

5. $12y^2 - 12xy - 5x^2 - 4$

6. Answers will vary.

7. Answers will vary. Possible answer: We used the zero property.

8. $5a^3 - 3a^2b^2 + 10b^3$

9. $10x^2 - y^2 - 15xy + 4$

10. $16c^3 - 8a^3 + 3ac^2 - 7$

Station 2

Students will be given 20 blue algebra tiles, 20 red algebra tiles, 20 green algebra tiles, and 20 yellow algebra tiles. Students will work together to model polynomials with algebra tiles. Then they will subtract polynomials using the algebra tiles.

Answers

1. $5x^2 + 5xy + 4y^2$

2. Answers will vary.

3. $3x^2, 2xy, 2y^2$

4. $-5x^2 - 5xy - 4y^2$

5. No, because the sign of the terms in the second polynomial changes to the opposite sign.

6. $7x^2 + 5xy + 11y^2$

7. Answers will vary. Possible answer: We matched like terms and then performed subtraction.

8. Answers will vary. Possible answer: We added negative terms because subtracting a negative number is the same as adding a positive of that number.

9. $-2a^4 - 4a^2b^2 + 6b^3 + 6$

10. $6c^2 - 6bc - 18$

Station 3

Students will be given a number cube. Students will use the number cube to populate coefficients of polynomials. Then they will multiply polynomials using the distributive property.

Answers

1. Answers will vary. Possible answer: $2x$ and $(3x + y - 2)$

2. distributive property

3. Answers will vary. Possible answer: $\begin{aligned} &2x(3x + y - 2) \\ &6x^2 + 2xy - 4x \end{aligned}$

4. Answers will vary. Possible answer: $-3x^2$ and $(-4x + 7xy - 8)$

5. Answers will vary. Possible answer: $12x^3 - 21x^3y + 24x^2$

6. It changed to the opposite sign because we multiplied each term by -1.

7. $(x + 3)$ and $(x - 4)$

8. distributive property

9. $(x + 3)(x - 4)$

$x^2 - 4x + 3x - 12$

$x^2 - x - 12$

10. We combined $-4x$ and $3x$.

Station 4

Students will be given six index cards with the following polynomials written on them:

$3xy^2$; $18x^2 - 7x + 4$; $33xy^5 - 3x^2y^2 - 21xy^2$; $2x$; $-3y^2$; $-24y^5 + 6y^3 - 12$

Students will work together to match polynomials and monomials that when divided by each other yield a specific quotient. Then students will perform synthetic division to divide a polynomial by a binomial.

Answers

1. $\dfrac{18x^2 - 7x + 4}{2x} = 9x - \dfrac{7}{2} + \dfrac{2}{x}$

2. $\dfrac{-24y^5 + 6y^3 - 12}{-3y^2} = 8y^3 - 2y + \dfrac{4}{y^2}$

3. $\dfrac{33xy^5 - 3x^2y^2 - 21xy^2}{3xy^2} = 11y^3 - x - 7$

4. Answers will vary. Possible answer: We divided each term by the monomial using the quotient rule for exponents.

5. No, because you have to divide by a binomial instead of a monomial.

6. 3; 3

7. Find the solution of the binomial, which is 1. Write this in the left hand box. Write the coefficients of the variables in a row. Bring down the first coefficient. Multiply this coefficient by 1. Add this product to the second coefficient. Repeat this process through all the coefficients. The last number is the remainder.

8. $2x^3 + 6x^2 + 7x + \dfrac{4}{x - 1}$

9. $4x^3 + 5x^2 + 15x + 31 + \dfrac{57}{x - 2}$

Materials List/Setup

Station 1	20 blue algebra tiles; 20 red algebra tiles; 20 green algebra tiles; 20 yellow algebra tiles
Station 2	20 blue algebra tiles; 20 red algebra tiles; 20 green algebra tiles; 20 yellow algebra tiles
Station 3	number cube
Station 4	six index cards with the following polynomials written on them:

$3xy^2$; $18x^2 - 7x + 4$; $33xy^5 - 3x^2y^2 - 21xy^2$; $2x$; $-3y^2$; $-24y^5 + 6y^3 - 12$

Discussion Guide

To support students in reflecting on the activities and to gather some formative information about student learning, use the following prompts to facilitate a class discussion to "debrief" the station activities.

Prompts/Questions

1. How do you add polynomials?

2. How do you subtract polynomials?

3. What happens to the exponents of the variables when you add or subtract polynomials?

4. How do you multiply polynomials?

5. How do you deal with the exponents of the variables when multiplying polynomials?

6. How do you divide a polynomial by a monomial?

7. How do you divide a polynomial by a binomial?

8. How do you deal with the exponents of the variables when dividing polynomials?

Think, Pair, Share

Have students jot down their own responses to questions, then discuss with a partner (who was not in their station group), and then discuss as a whole class.

Suggested Appropriate Responses

1. Add like terms.

2. Subtract like terms of the second polynomial from like terms in the first polynomial.

3. Exponents remain the same in addition and subtraction of polynomials.

4. Multiply each term in the first polynomial by each term in the second polynomial using the distributive property.

5. Use the product rule on the exponents.

6. Divide each term in the polynomial by the monomial.

7. Use synthetic division.

8. Use the quotient rule on the exponents.

Possible Misunderstandings/Mistakes

- Incorrectly adding exponents when adding polynomials
- Incorrectly subtracting exponents when subtracting polynomials
- Not using the product rule on exponents when multiplying polynomials
- Not using the quotient rule on exponents when dividing polynomials
- Not using synthetic division when dividing by a binomial
- Not realizing that the last number in synthetic division is the remainder

Arithmetic with Polynomials and Rational Expressions
Set 1: Operations with Polynomials

Station 1

At this station, you will find 20 blue algebra tiles, 20 red algebra tiles, 20 green algebra tiles, and 20 yellow algebra tiles. Work as a group to model each polynomial by placing the tiles next to the polynomials. Then find the sum.

- Use the blue algebra tiles to model the x^2 term.

- Use the red algebra tiles to represent the xy term.

- Use the green algebra tiles to represent the y^2 term.

- Use the yellow algebra tiles to represent the constant.

1. Given:

$$3x^2 + 2xy + 2y^2$$
$$+ \quad 5x^2 - xy + 3y^2$$

Answer: _____

2. How did you use the algebra tiles to model the problem?

3. How did you model the $-xy$ term?

4. What property did you use on the xy terms? _____

5. Model the following problem using the algebra tiles. Show your work.

$$(4y^2 - 12xy + 5x^2) + (-10x^2 + 8y^2 - 4)$$

continued

Answer: _____

Arithmetic with Polynomials and Rational Expressions
Set 1: Operations with Polynomials

6. How did you use the algebra tiles to model problem 5?

7. How did you deal with negative terms during addition?

Work together to add each polynomial. Show your work.

8. Given:

$$2a^3 + a^2b^2 + 3b^3$$
$$+ \quad 3a^3 - 4a^2b^2 + 7b^3$$

9. $-10xy - 3 + 2x^2 - 5y^2 + 4y^2 + 8x^2 - 5xy + 7$

10. $8c^3 + 3ac^2 + 4a^3 + 8c^3 - 12a^3 - 7$

Arithmetic with Polynomials and Rational Expressions
Set 1: Operations with Polynomials

Station 2

At this station, you will find 20 blue algebra tiles, 20 red algebra tiles, 20 green algebra tiles, and 20 yellow algebra tiles. Work as a group to model each polynomial by placing the tiles next to the polynomials. Then find the difference.

- Use the blue algebra tiles to model the x^2 term.
- Use the red algebra tiles to represent the xy term.
- Use the green algebra tiles to represent the y^2 term.
- Use the yellow algebra tiles to represent the constant.

1. Given:
$$8x^2 + 7xy + 6y^2$$
$$- (3x^2 + 2xy + 2y^2)$$

Answer: _____

2. How did you use the algebra tiles to model the problem?

3. What terms in the bottom polynomial does the subtraction sign apply to?

4. Find the difference:
$$3x^2 + 2xy + 2y^2$$
$$- (8x^2 + 7xy + 6y^2)$$

Answer: _____

continued

Arithmetic with Polynomials and Rational Expressions
Set 1: Operations with Polynomials

5. Is your answer from problem 1 the same as your answer from problem 4? Why or why not?

6. Model the subtraction problem below using the algebra tiles, then solve. Show your work.

$$2x^2 + 5y^2 + 9xy$$
$$- \quad (4xy - 5x^2 - 6y^2)$$
$$\overline{}$$

Answer: _____

7. How did you arrange the algebra tiles to model problem 6?

8. How did you deal with negative terms during subtraction?

9. Work together to subtract each polynomial. Show your work.

$$a^4 - a^2b^2 + 4b^3 + 8$$
$$- \quad (3a^4 + 3a^2b^2 - 2b^3 + 2)$$
$$\overline{}$$

10. Subtract $8c^2 + 2bc + 10$ from $-4bc + 14c^2 - 8$.

Arithmetic with Polynomials and Rational Expressions
Set 1: Operations with Polynomials

Station 3

At this station, you will find a number cube. As a group, roll the number cube. Write the result in the box below.

Given: [   ] $x(3x + y - 2)$

1. Identify the two polynomials above: _____

2. What property can you use to multiply these polynomials?

   _____

3. Multiply the polynomials. Show your work.

Given: $-$ [   ] $x^2(-4x + 7xy - 8)$

4. Identify the two polynomials above: _____

5. Multiply the polynomials. Show your work.

continued

Algebra I Station Activities for Common Core State Standards

6. What happened to the signs of each term of the polynomial in the parentheses? Explain your answer.

Given: $(x + 3)(x - 4)$

7. Identify the two polynomials above: _____

8. What method can you use to multiply these polynomials?

_____

9. Multiply the polynomials. Show your work.

10. What extra steps did you take when multiplying $(x + 3)(x - 4)$ versus $-\boxed{}x^2(-4x + 7xy - 8)$?

Arithmetic with Polynomials and Rational Expressions
Set 1: Operations with Polynomials

Station 4

At this station, you will find six index cards with the following polynomials written on them:

$$3xy^2;\ 18x^2 - 7x + 4;\ 33xy^5 - 3x^2y^2 - 21xy^2;\ 2x;\ -3y^2;\ -24y^5 + 6y^3 - 12$$

Shuffle the cards. Work as a group to match the polynomials that when divided yield each quotient below. (*Hint*: Place the monomials in the denominator.)

1. $9x - \dfrac{7}{2} + \dfrac{2}{x}$

2. $8y^3 - 2y + \dfrac{4}{y^2}$

3. $11y^3 - x - 7$

4. What strategy did you use for problems 1–3?

Given: $(2x^3 + 4x^2 + x - 3) \div (x - 1)$

5. Can you use the same strategy to divide the polynomials above as you did in problems 1–3? Why or why not?

6. What is the degree of $(2x^3 + 4x^2 + x - 3)$? _____

   This degree means the quotient will have _____ terms.

continued

Arithmetic with Polynomials and Rational Expressions
Set 1: Operations with Polynomials

Follow these steps to use synthetic division to find the quotient of $(2x^3 + 4x^2 + x - 3) \div (x - 1)$.

Step 1: Set the binomial equal to zero and solve for x. Show your work.

Step 2: Use your answer from Step 1 and write it in the first box on the left in the illustration below.

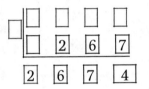

Step 3: Write the coefficients of each term in order of left to right in the top row of boxes in the illustration under Step 2.

Step 4: The first coefficient (in this case, 2 of 2 • 1) is always written in the box underneath it.

Step 5: The boxes that are already filled in show synthetic division.

7. Derive the process of synthetic division based on the example above.

Step 6: The last number in the bottom row, 4, is known as the **remainder** and can be written as $\dfrac{4}{x-1}$.

8. Based on steps 1–6, what is the answer to $(2x^3 + 4x^2 + x - 3) \div (x - 1)$ using synthetic division? _____

9. Use synthetic division to find $(4x^4 - 3x^3 + 5x^2 + x - 5) \div (x - 2)$. Show your work.

Interpreting Functions

Set 1: Relations Versus Functions/Domain and Range

Goal: To provide opportunities for students to develop concepts and skills related to using function notation, domain, range, relations, and functions

Common Core Standards

Functions: Interpreting Functions

Understand the concept of a function and use function notation.

F-IF.1. Understand that a function from one set (called the domain) to another set (called the range) assigns to each element of the domain exactly one element of the range. If f is a function and x is an element of its domain, then $f(x)$ denotes the output of f corresponding to the input x. The graph of f *is the graph of the equation* $y = f(x)$.

F-IF.2. Use function notation, evaluate functions for inputs in their domains, and interpret statements that use function notation in terms of a context.

Functions: Building Functions

Build a function that models a relationship between two quantities.

F-BF.1. Write a function that describes a relationship between two quantities.★

a. Determine an explicit expression, a recursive process, or steps for calculation from a context.

Student Activities Overview and Answer Key

Station 1

Students will be given eight index cards with functions and function answers on them. They will match the functions with the appropriate function answers. Then they will evaluate functions.

Answers

1. $f(x) = 2x$ with $f(3) = 6$; $f(x) = -3t + 7$ with $f(3) = -2$; $f(x) = x^2$ with $f(3) = 9$; $f(x) = \dfrac{2}{3}x$ with $f(3) = 2$.

2. $f(x + 3) = x + 8$

3. $f(t - 4) = t^2 - 8t + 16$ or $(t - 4)(t - 4)$ or $(t - 4)^2$

4. $f(s + 4) = \dfrac{1}{5}s + \dfrac{4}{5}$ or $\dfrac{(s + 4)}{5}$

Station 2

Students will use a ruler to perform the vertical line test on graphs of relations. They will determine if the relation is a function. They will construct a graph that is a function. Then they will determine if a relation is a function by analyzing coordinate pairs.

Answers

1. Yes; the vertical line test holds.

2. No; the vertical line test does not hold.

3. Yes; the vertical line test holds; I used the vertical line test, which says if any vertical line passes through a graph at more than one point, then the graph is not the graph of a function.

4. Answers will vary. Verify that the vertical line test holds.

5. Not a function because the element 3 in the domain has two assigned elements in the range. (3, 1) and (3, 6)

6. Yes, it is a function.

Station 3

Students will be given a calculator to help them solve a real-world linear function. They will write and solve a linear function based on two data points.

Answers

1. (100, 19), (250, 17)

2. slope $= -\dfrac{1}{75}$

3. Use the point (100, 19).

$$y - 19 = -\frac{1}{75}(x - 100)$$

$$y = -\frac{x}{75} + \frac{61}{3} \text{ or } f(x) = -\frac{x}{75} + \frac{61}{3}$$

4. $f(500) = -\dfrac{1(500)}{75} + \dfrac{61}{3} = \$13.67$

5. $f(60) = -\dfrac{1(60)}{75} + \dfrac{61}{3} = \$19.53$

Station 4

Students will be given a number cube. They roll the number cube to populate a relation. They find the domain and range of the relation and determine if it is a function. Then for given relations, they determine the domain, range, and whether or not it is a function.

Answers

1. Answers will vary; verify that the domain includes the x-values.

2. Answers will vary; verify that the range includes the y-values.

3. Answers will vary; a function is a relation in which each x input has only one y output.

4. Domain: {–1, 2, 3, 4}; range: {2, 5, 10}; yes, it is a function.

5. Domain: {3, 7, 10}; range: {2, 5, 7}; not a function because there are two y-values for x = 10: (10, 7) and (10, 5).

6. Domain: {–14, 14, 15, 17}; range: {–9, 8, 17}; yes, it is a function.

Materials List/Setup

Station 1   eight index cards with the following functions and answers written on them:

$$f(x) = 2x; f(x) = -3t + 7; f(x) = x^2; \ f(x) = \frac{2}{3}x; f(3) = 6; f(3) = 9; f(3) = -2; f(3) = 2$$

Station 2   ruler; graph paper

Station 3   calculator

Station 4   number cube

Discussion Guide

To support students in reflecting on the activities and to gather some formative information about student learning, use the following prompts to facilitate a class discussion to "debrief" the station activities.

Prompts/Questions

1. How do you evaluate a function $f(x)$ when given a value for x?

2. What is the vertical line test for a function?

3. What is the general formula of a linear function? How does this relate to a linear equation?

4. How do you find the domain and range of a relation?

5. How can you determine whether or not a relation is a function?

Think, Pair, Share

Have students jot down their own responses to questions, then discuss with a partner (who was not in their station group), and then discuss as a whole class.

Suggested Appropriate Responses

1. Plug the value of x into the function to solve for $f(x)$.

2. The vertical line test says if any vertical line passes through a graph at more than one point, then the graph is not the graph of a function.

3. $f(x) = mx + b$ where m and b are real numbers and $m \neq 0$. This is the same as $y = mx + b$.

4. The domain is the x-values. The range is the y-values.

5. For every x input value, there must only be one y output value assigned to it.

Possible Misunderstandings/Mistakes

- Mixing up the domain and range
- Incorrectly thinking that in a function each y-value must have a unique x-value assigned to it
- Not keeping track of variables plugged into a function
- Using a horizontal line test instead of a vertical line test to determine if a relation is a function

Interpreting Functions
Set 1: Relations Versus Functions/Domain and Range

Station 1

You will be given eight index cards with the following functions and answers written on them:

$$f(x) = 2x; f(x) = -3t + 7; f(x) = x^2; \ f(x) = \frac{2}{3}x; f(3) = 6; f(3) = 9; f(3) = -2; f(3) = 2$$

1. Work together to match the appropriate function with each answer. Write your matches below.

Solve. Show your work.

2. Let $f(x) = x + 5$. What is $f(x + 3)$?

3. Let $f(t) = t^2$. What is $f(t - 4)$?

4. Let $f(s) = \frac{1}{5}s$. What is $f(s + 4)$?

Interpreting Functions
Set 1: Relations Versus Functions/Domain and Range

Station 2

You will be given a ruler and graph paper. As a group, use your ruler to determine whether or not each relation below is a function. Beside each graph, write your answer and reasoning.

1.

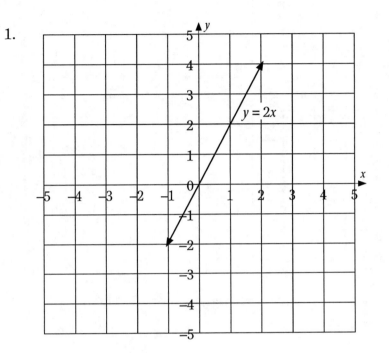

$y = 2x$

2.

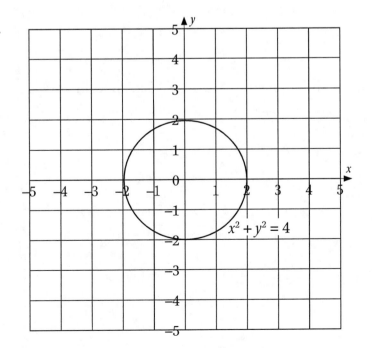

$x^2 + y^2 = 4$

continued

Interpreting Functions
Set 1: Relations Versus Functions/Domain and Range

3.

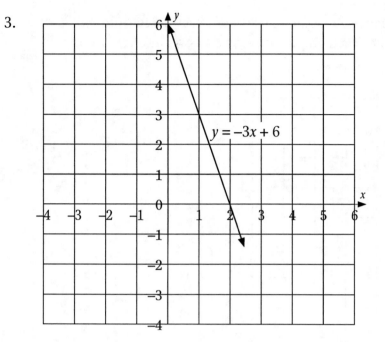

$y = -3x + 6$

How did you use your ruler to determine whether each relation was a function?

4. Use your ruler and graph paper to sketch a function. Use the vertical line test to verify that it is a function.

For the relations below, determine whether or not they are functions. Explain your answer.

5. {(2, 5), (3, 1), (1, 4), (3, 6)}

6. {(1, 1), (2, 1), (3, 2)}

Interpreting Functions
Set 1: Relations Versus Functions/Domain and Range

Station 3

A function f is linear if $f(x) = mx + b$, where m and b are real numbers and $m \neq 0$.

Use this information and the problem scenario below to answer the following questions. You may use a calculator.

> The cost of a sweatshirt is linearly related to the number of sweatshirts ordered. If you buy 100 sweatshirts, then the cost per sweatshirt is $19. However, if you buy 250 sweatshirts, then the cost per sweatshirt is only $17.

1. You are given two points in the function. If x represents the number of sweatshirts and y represents the cost per sweatshirt, write the two ordered pairs represented in the problem scenario above.

2. What is the slope of the function?

3. Find a function which relates the number of sweatshirts and the cost per sweatshirt. Show your work in the space below.

4. What would the cost per sweatshirt be for 500 sweatshirts? Explain.

5. What would the cost per sweatshirt be for 60 sweatshirts? Explain.

© 2011 Walch Education

Interpreting Functions
Set 1: Relations Versus Functions/Domain and Range

Station 4

You will be given a number cube. As a group, roll the number cube and write the result in the first box. Repeat this process until all the boxes contain a number.

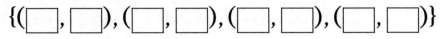

1. What is the domain of this relation?

2. What is the range of this relation?

3. Is this relation a function? Why or why not?

For problems 4–6, state the domain, range, and whether or not the relation is a function. Include your reasoning.

4. {(2, 5), (3, 10), (–1, 2), (4, 5)}

5. {(10, 7), (3, 7), (10, 5), (7, 2)}

6. {(–14, 8), (17, 8), (14, –9), (15, 17)}

Interpreting Functions

Set 2: Graphing Quadratic Equations

Goal: To provide opportunities for students to develop concepts and skills related to graphing quadratic equations and functions

Common Core Standards

Algebra: Reasoning with Equations and Inequalities

Represent and solve equations and inequalities graphically.

A-REI.10. Understand that the graph of an equation in two variables is the set of all its solutions plotted in the coordinate plane, often forming a curve (which could be a line).

Functions: Interpreting Functions

Analyze functions using different representations.

F-IF.7. Graph functions expressed symbolically and show key features of the graph, by hand in simple cases and using technology for more complicated cases.★

  a. Graph linear and quadratic functions and show intercepts, maxima, and minima.

Student Activities Overview and Answer Key

Station 1

Students will be given graph paper and a ruler. Students will derive how to find the vertex of the graph of a quadratic function. Then they will find the x-intercept and y-intercept of the function. They will graph the quadratic function and describe its shape.

Answers

1. $f(x) = x^2 + 6x + 9$

2. $a = 1; b = 6;$ and $c = 9$

3. $\dfrac{-b}{2a}$

4. $f\left(\dfrac{-b}{2a}\right)$

5. $\left(\dfrac{-b}{2a}, f\left(\dfrac{-b}{2a}\right)\right); (-3, 0)$

6. Set $y = f(x) = 0$ and solve for x by factoring the function.

7. Set $x = 0$ and solve for y.

8. $(-3, 0)$ and $(0, 9)$

9.

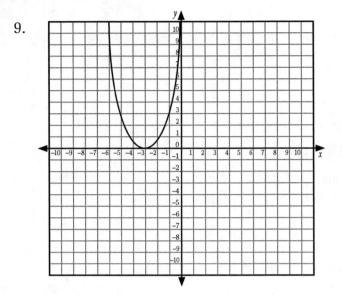

10. Parabola; the square term gives two x-values for each y-value due to the nature of square roots.

Station 2

Students will be given a graphing calculator. Students will use the graphing calculator to graph two quadratic functions. They will describe and analyze the characteristics of both graphs, including the vertex.

Answers

1. parabola

2. upward

3. x^2, because it has a positive coefficient

4. table of (x, y) values that satisfy the function

5. Find the y-value where $x = 0$. This point is the vertex; $(0, 4)$

6. parabola

7. upward

8. x^2, because it has a positive coefficient

9. table of (x, y) values for both functions

10. Find the y-value where $x = 0$. This point is the vertex; $(0, -4)$

11. The value of the constant determines the vertex of the graph.

Station 3

Students will be given a graphing calculator. Students will use the graphing calculator to graph three quadratic functions. The quadratic functions have the same vertex, but varying widths. Students will describe the relationships between the widths of the parabolas.

Answers

1. vertex at (0, 0) because there is no constant

2. $y = x^2$; because the coefficient is 1 versus 3. A smaller coefficient yields a wider parabola.

3. table of x- and y-values for both graphs

4. $Y_2 = 3Y_1$; it shows that the y-value for each x-value of Y_2 is three times as large as Y_1.

5. because it has a smaller coefficient

6. table of x- and y-values for all three graphs

7. $Y_3 = 1/2(Y_1)$; it shows that the y-value for each x-value of Y_3 is half the size of Y_1.

8. $Y_3 = 1/6(Y_2)$; it shows that the y-value for each x-value of Y_3 is one-sixth the size of Y_2.

Station 4

Students will be given graph paper and a ruler. Students will analyze and graph quadratic equations using vertices, x-intercepts, and a table of values. They will determine why certain parabolas open upward while others open downward.

Answers

1. $f(x) = x^2 - x - 6$ and $f(x) = -x^2 + x - 6$

$a = 1$	$a = -1$
$b = -1$	$b = 1$
$c = -6$	$c = -6$

2. For $f(x) = x^2 - x - 6$, the vertex is $(1/2, -23/4)$.

   For $f(x) = -x^2 + x - 6$, the vertex is $(1/2, -25/4)$.

3. $(3, 0), (-2, 0)$

4.

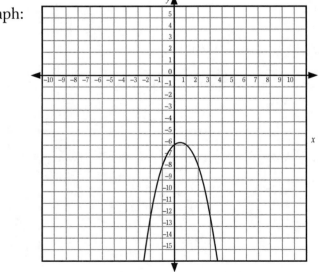

5. upward, because the coefficient of x^2 is positive

6. Table:

x	$y = f(x)$
−4	−26
0	−6
4	−18

Graph:

7. downward, because the coefficient of x^2 is negative

8. no, because the parabola opens downward

Materials List/Setup

Station 1   graph paper; ruler

Station 2   graphing calculator

Station 3   graphing calculator

Station 4   graph paper; ruler

Discussion Guide

To support students in reflecting on the activities and to gather some formative information about student learning, use the following prompts to facilitate a class discussion to "debrief" the station activities.

Prompts/Questions

1. What does the graph of a quadratic equation look like?

2. Why does the graph have this shape?

3. How do you find the vertex of a quadratic equation?

4. Does a quadratic equation with a term of x^2 open upward or downward? Why?

5. Does a quadratic equation with a term of $-x^2$ open upward or downward? Why?

6. How can you find the x-intercepts of a quadratic equation?

Think, Pair, Share

Have students jot down their own responses to questions, then discuss with a partner (who was not in their station group), and then discuss as a whole class.

Suggested Appropriate Responses

1. The graph of a quadratic equation looks like a parabola.

2. There are two x-values for each y-value because of the x^2 term.

3. vertex $= \left(\dfrac{-b}{2a}, f\left(\dfrac{-b}{2a} \right) \right)$

4. upward, because the coefficient is positive

5. downward, because the coefficient is negative

6. Factor the equation to find the x-intercepts.

Possible Misunderstandings/Mistakes

- Not realizing that graphs of quadratic equations are parabolas

- Not using enough data points to construct a parabola for the graph of the quadratic equation

- Ignoring the sign of the coefficient of the x^2 term when constructing a parabola that opens upward or downward

Interpreting Functions
Set 2: Graphing Quadratic Equations

Station 1

At this station, you will find graph paper and a ruler. Work together to graph the following quadratic equation:

$$y = x^2 + 6x + 9$$

1. Write this quadratic equation as a quadratic function. _____

2. What are the values of a, b, and c in the quadratic function?

   $a = $ _____

   $b = $ _____

   $c = $ _____

To graph the function, you need the vertex, x-intercept, and y-intercept.

3. If the x-value of the vertex is found by $x = \dfrac{-6}{2(1)} = -3$, then write this x calculation using the general terms a, b, and/or c.

   _____

4. If the y-value of the vertex is found by $y = f\left(\dfrac{-6}{2(1)}\right) = f(-3) = 0$, then write this y calculation using the general terms a, b, and/or c.

   _____

5. Based on problems 3 and 4, how can you find the vertex of the graph for $f(x) = ax^2 + bx + c$?

   What is the vertex of the quadratic function $x^2 + 6x + 9 = 0$? _____

continued

Interpreting Functions
Set 2: Graphing Quadratic Equations

6. How do you find the *x*-intercept of a function? (*Hint*: $y = f(x)$)

7. How do you find the *y*-intercept of a function?

8. What are the intercepts for $y = x^2 + 6x + 9$? _____

9. On your graph paper, graph the function using the vertex, *x*-intercept, and *y*-intercept.

10. What shape is the graph? Why do you think the graph has this shape?

Interpreting Functions
Set 2: Graphing Quadratic Equations

Station 2

At this station, you will find a graphing calculator. As a group, follow these steps to graph
$y = x^2 + 4$ and $y = x^2 - 4$.

Step 1: Hit the "Y =" key. At Y1 =, type in $x\textasciicircum2 + 4$.

Step 2: Hit the "GRAPH" key.

---

1. What shape is the graph? _____

2. Does the graph open upward or downward? _____

3. Which term do you think makes the graph open upward or downward? Explain your reasoning.

Step 3: Hit the "2nd" key then the "GRAPH" key.

---

4. What information does your calculator show?

5. How can you use this information to find the vertex of the graph?

   What is the vertex of the graph? _____

continued

Interpreting Functions
Set 2: Graphing Quadratic Equations

Step 4: Hit "Y =". At Y2 =, type in $x^2 - 4$.

Step 5: Hit the "GRAPH" key.

---

6. What shape is the graph? _____

7. Does the graph open upward or downward? _____

8. Which term do you think makes the graph open upward or downward? Explain your reasoning.

Step 6: Hit the "2nd" key then the "GRAPH" key.

---

9. What information does your calculator show?

10. How can you use this information to find the vertex of the graph of $y = x^2 - 4$?

    What is the vertex of $y = x^2 - 4$? _____

11. Why do the graphs for $y = x^2 + 4$ and $y = x^2 - 4$ have different vertices?

Interpreting Functions
Set 2: Graphing Quadratic Equations

Station 3

At this station, you will find a graphing calculator. As a group, follow these steps to graph $y = x^2$, $y = 3x^2$, and $y = \frac{1}{2}x^2$.

Step 1: Hit the "Y =" key.

At Y1 =, type in x^2.

At Y2 =, type in $3x$^2.

Step 2: Hit the "GRAPH" key.

---

1. Why do both graphs have the same vertex?

2. Which graph is wider, $y = x^2$ or $y = 3x^2$? _____

   Why is one graph wider than the other?

Step 3: Hit the "2nd" key then the "GRAPH" key.

---

3. What information does your calculator show?

4. What is the relationship between Y1 and Y2 in the table?

   How does this relationship relate to $y = x^2$ and $y = 3x^2$?

continued

Interpreting Functions
Set 2: Graphing Quadratic Equations

Step 4: At Y3 =, type in $0.5x^2$.

Step 5: Hit the "GRAPH" key.

---

5. Why is the graph of $y = 0.5x^2$ wider than $y = x^2$ and $y = 3x^2$?

Step 6: Hit the "2nd" key then the "GRAPH" key.

---

6. What information does your calculator show?

7. What is the relationship between Y1 and Y3 in the table?

   How does this relationship relate to $y = x^2$ and $y = 0.5x^2$?

8. What is the relationship between Y2 and Y3 in the table?

   How does this relationship relate to $y = 3x^2$ and $y = 0.5x^2$?

Algebra I Station Activities for Common Core State Standards

Interpreting Functions
Set 2: Graphing Quadratic Equations

Station 4

At this station, you will find graph paper and a ruler. Work together to graph the following quadratic equations:

$$f(x) = x^2 - x - 6 \text{ and } f(x) = -x^2 + x - 6$$

1. What are the values of a, b, and c in each quadratic function?

   $f(x) = x^2 - x - 6$ $f(x) = -x^2 + x - 6$

   $a =$ _____ $a =$ _____

   $b =$ _____ $b =$ _____

   $c =$ _____ $c =$ _____

2. Use the information in problem 1 to find the vertex $\left(\dfrac{-b}{2a}, f\left(\dfrac{-b}{2a} \right) \right)$ for each function. Show your work.

3. Find the x-intercepts of $f(x) = x^2 - x - 6$ using factoring. Show your work.

4. On your graph paper, graph $f(x) = x^2 - x - 6$ using its vertex and x-intercepts.

5. Does the parabola open upward or downward? Explain your answer.

continued

Interpreting Functions
Set 2: Graphing Quadratic Equations

6. Fill out the table below to help you graph $f(x) = -x^2 + x - 6$.

x	$y = f(x)$
-4	
0	
4	

Graph $f(x) = -x^2 + x - 6$ on your graph paper.

7. Does the graph open upward or downward? Explain your answer.

8. Will the graph of $f(x) = -x^2 + x - 6$ have x-intercepts? Why or why not?

Interpreting Functions

Set 3: Comparing Linear, Exponential, Quadratic, and Absolute Value Models

Goal: To provide opportunities for students to develop concepts and skills related to comparing linear, exponential, and quadratic models and absolute value

Common Core Standards

Algebra: Reasoning with Equations and Inequalities

Represent and solve equations and inequalities graphically.

A-REI.10. Understand that the graph of an equation in two variables is the set of all its solutions plotted in the coordinate plane, often forming a curve (which could be a line).

Functions: Interpreting Functions

Analyze functions using different representations.

F-IF.7. Graph functions expressed symbolically and show key features of the graph, by hand in simple cases and using technology for more complicated cases.★

   a. Graph linear and quadratic functions and show intercepts, maxima, and minima.

   b. Graph square root, cube root, and piecewise-defined functions, including step functions and absolute value functions.

   c. Graph polynomial functions, identifying zeros when suitable factorizations are available, and showing end behavior.

   d. (+) Graph rational functions, identifying zeros and asymptotes when suitable factorizations are available, and showing end behavior.

   e. Graph exponential and logarithmic functions, showing intercepts and end behavior, and trigonometric functions, showing period, midline, and amplitude.

Student Activities Overview and Answer Key

Station 1

Students will be given a linear equation and asked to generate a table of values and the graph. Then students will examine the equation, table of values, and graph for defining characteristics of linear equations.

Answers

1. Answers will vary. Sample answer:

x	y
−2	1
−1	2
0	3
1	4
2	5

2. x-intercept: −3; y-intercept: 3

3.

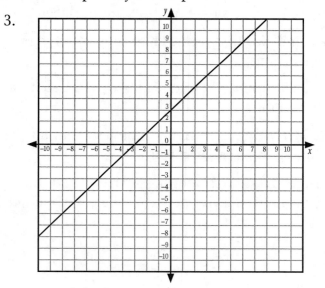

4. Answers will vary. Sample answer: power of 1 on x

5. Answers will vary. Sample answer: Constant rate of change in y with respect to x. When x increases by 1, so does y.

6. Answers will vary. Sample answer: straight line

Station 2

Students will be given a quadratic equation and asked to generate a table of values and the graph. Then students will examine the equation, table of values, and graph for defining characteristics of quadratic equations.

Answers

1. Answers will vary. Sample answer:

x	y
−2	0
−1	−3
0	−4
1	−3
2	0

2. x-intercepts: −2 and 2; y-intercept: −4

3.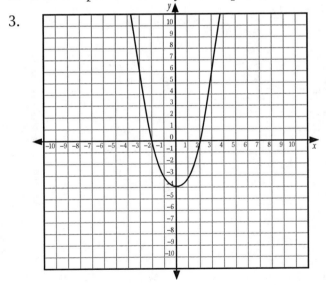

4. Answers will vary. Sample answer: power of 2 on x

5. Answers will vary. Sample answer: variable rate of change in y with respect to x. When x increases by 1, y does not change at a constant rate.

6. Answers will vary. Sample answer: U-shape or parabola

Station 3

Students will be given an absolute value equation and asked to generate a table of values and the graph. Then students will examine the equation, table of values, and graph for defining characteristics of absolute value equations.

Answers

1. Answers will vary. Sample answer:

x	y
−2	3
−1	2
0	1
1	2
2	3

2. *x*-intercepts: none; *y*-intercept: 1

3.

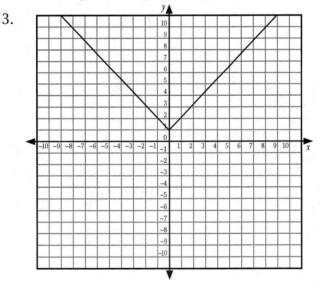

4. Answers will vary. Sample answer: absolute value bars around *x* and power of 1 on *x*

5. Answers will vary. Sample answer: constant rate of change for each "leg"

6. Answers will vary. Sample answer: V-shape

Interpreting Functions
Set 3: Comparing Linear, Exponential, Quadratic, and Absolute Value Models

Station 4

Students will be given an exponential equation and asked to generate a table of values and the graph. Then students will examine the equation, table of values, and graph for defining characteristics of exponential value equations.

Answers

1. Answers will vary. Sample answer:

x	y
–2	1/4
–1	1/2
0	1
1	2
2	4
3	8
4	16

2. x-intercepts: none; y-intercept: 1

3.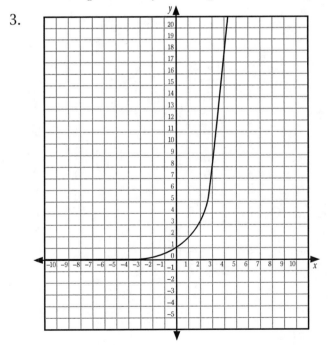

4. Answers will vary. Sample answer: variable in exponent

5. Answers will vary. Sample answer: It "grows" quickly.

6. Answers will vary. Sample answer: It rises to the right and levels off toward the left.

Algebra I Station Activities for Common Core State Standards

Materials List/Setup

Station 1	ruler
Station 2	none
Station 3	none
Station 4	none

Discussion Guide

To support students in reflecting on the activities and to gather some formative information about student learning, use the following prompts to facilitate a class discussion to "debrief" the station activities.

Prompts/Questions

1. How do you determine if an equation is linear? Quadratic? Exponential? Absolute value?

2. What is the general shape of the graph of a linear equation?

3. What is the general shape of the graph of an exponential function?

4. What is the general shape of the graph of a quadratic function?

5. What is the general shape of the graph of a linear absolute value equation?

6. How are your linear and absolute value equations similar? How are they different?

7. How are linear absolute value equations and quadratic equations similar? How are they different?

Think, Pair, Share

Have students jot down their own responses to questions, then discuss with a partner (who was not in their station group), and then discuss as a whole class.

Suggested Appropriate Responses

1. An equation is linear if the variables are to the power of 1, the variables are not multiplied together, and the variable is not in the denominator. A quadratic function has a squared variable. An exponential equation has a variable in the exponent. An absolute value equation has absolute value symbols around the variable.

2. a line

3. a curve that extends toward infinity on one side and approaches the x-axis on the other side

4. a parabola or U-shape

5. a V-shape

6. They both have constant rates of change. The absolute value equation has two opposite rates of change, while a linear equation has one constant rate of change.

7. Absolute value equations and quadratic equations both have a line of symmetry. Linear absolute value equations have two constant rates of change, while quadratic equations have variable rates of change.

Possible Misunderstandings/Mistakes

- Not generating the table of values correctly
- Plotting points incorrectly
- Miscalculating the x- and y-intercepts

Interpreting Functions
Set 3: Comparing Linear, Exponential, Quadratic, and Absolute Value Models

Station 1

You will work with a linear equation at this station.

Use the linear equation below for the following problems.

$$y = x + 3$$

1. Create a table of values for your equation.

x	y

2. Find the x- and y-intercepts.

3. Graph your equation below.

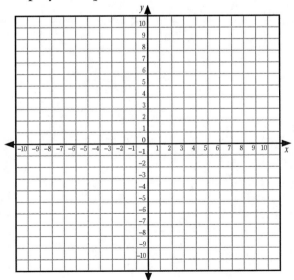

continued

Interpreting Functions
Set 3: Comparing Linear, Exponential, Quadratic, and Absolute Value Models

4. Looking at the equation, what are some defining characteristics of a linear equation?

5. Looking at the table of values, what are some defining characteristics of a linear equation's table of values?

6. Looking at the graph, what are some defining characteristics of a linear equation's graph?

Interpreting Functions
Set 3: Comparing Linear, Exponential, Quadratic, and Absolute Value Models

Station 2

You will work with a quadratic equation at this station.

Use the quadratic equation below for the following problems.

$$y = x^2 - 4$$

1. Create a table of values for your equation.

x	y

2. Find the x- and y-intercepts.

3. Graph your equation below.

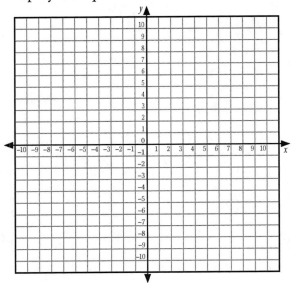

continued

Interpreting Functions
Set 3: Comparing Linear, Exponential, Quadratic, and Absolute Value Models

4. Looking at the equation, what are some defining characteristics of a quadratic equation?

5. Looking at the table of values, what are some defining characteristics of a quadratic equation's table of values?

6. Looking at the graph, what are some defining characteristics of a quadratic equation's graph?

Interpreting Functions

Set 3: Comparing Linear, Exponential, Quadratic, and Absolute Value Models

Station 3

You will work with an absolute value equation at this station.

Use the absolute value equation below for the following problems.

$$y = |x| + 1$$

1. Create a table of values for your equation.

x	y

2. Find the *x*- and *y*-intercepts.

3. Graph your equation below.

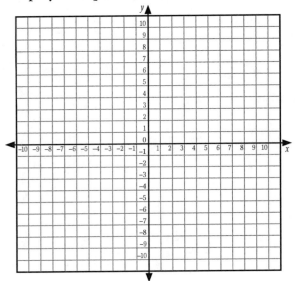

Algebra I Station Activities for Common Core State Standards

Interpreting Functions
Set 3: Comparing Linear, Exponential, Quadratic, and Absolute Value Models

4.   Looking at the equation, what are some defining characteristics of an absolute value equation?

5.   Looking at the table of values, what are some defining characteristics of an absolute value equation's table of values?

6.   Looking at the graph, what are some defining characteristics of an absolute value equation's graph?

Interpreting Functions
Set 3: Comparing Linear, Exponential, Quadratic, and Absolute Value Models

Station 4

You will work with an exponential equation at this station.

Use the exponential equation below for the following problems.

$$y = 2^x$$

1. Create a table of values for your equation.

x	y

2. Find the *x*- and *y*-intercepts.

3. Graph your equation below.

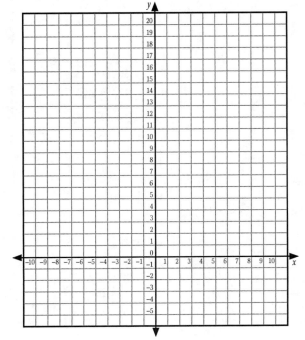

continued

Interpreting Functions
Set 3: Comparing Linear, Exponential, Quadratic, and Absolute Value Models

4. Looking at the equation, what are some defining characteristics of an exponential equation?

5. Looking at the table of values, what are some defining characteristics of an exponential equation's table of values?

6. Looking at the graph, what are some defining characteristics of an exponential equation's graph?

Statistics and Probability

Set 1: Line of Best Fit

Goal: To provide opportunities for students to develop concepts and skills related to creating and analyzing scatter plots and lines of best fit to represent a real-world situation

Common Core Standards

Statistics and Probability: Interpreting Categorical and Quantitative Data

Summarize, represent, and interpret data on two categorical and quantitative variables.

S-ID.6. Represent data on two quantitative variables on a scatter plot, and describe how the variables are related.

    a. Fit a function to the data; use functions fitted to data to solve problems in the context of the data. Use given functions or choose a function suggested by the context. Emphasize linear, quadratic, and exponential models.

    b. Informally assess the fit of a function by plotting and analyzing residuals.

    c. Fit a linear function for a scatter plot that suggests a linear association.

Interpret linear models.

S-ID.7. Interpret the slope (rate of change) and the intercept (constant term) of a linear model in the context of the data.

Student Activities Overview and Answer Key

Station 1

Students will be given graph paper and a ruler to help them create a scatter plot. Then they will analyze the scatter plot to determine the correlation between the data and describe the slope.

Answers

1. Graph:

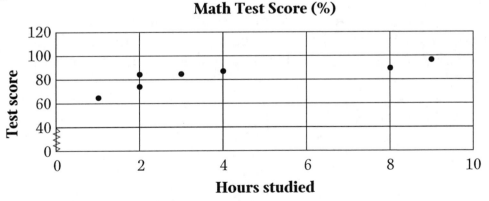

The graph is a scatter plot.

2. The test scores increase as the amount of time she studies for each test increases.

3. Positive correlation; the longer she studied for the test, the higher her test score; positive slope because the line increases from left to right.

4. No, you can't apply this observation to other data sets. She may need to spend more or less time studying in English or history in order to earn the same test scores.

5. Answers will vary; verify that the scatter plot depicts a positive correlation.

6. Answers will vary; verify that the scatter plot depicts a negative correlation.

7. Answers will vary; verify that the scatter plot depicts no correlation.

Station 2

Students will be given graph paper and a ruler. They will find the line of best fit and describe its slope for the set of data. Then they will create a scatter plot and line of best fit that represents a real-world situation.

Answers

1. Line D; Line D is the closest to the most number of data points.

2. Answers will vary for line of best fit; check each group's graph paper. Slope is positive.

3. Answers will vary for line of best fit; check each group's graph paper. Slope is negative.

4. Answers will vary for line of best fit; check each group's graph paper. Slope is zero or close to zero.

5. Answers will vary; verify that students created the scatter plot and line of best fit correctly based on their set of data points.

Station 3

Students will be given a ruler and graph paper. They will construct a scatter plot and line of best fit for a real-world situation. They will find the slope of the line and describe how it represents the data set. Then they will write an equation for the line of best fit in slope-intercept form.

Answers

1.

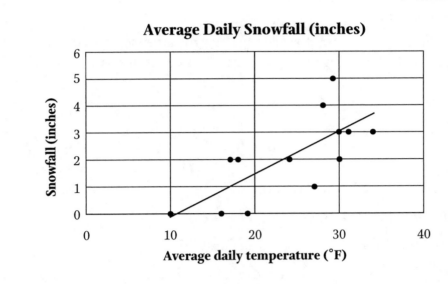

Average Daily Snowfall (inches)

(30, 3) and (10, 0) because they are closest to the line of best fit.

2. $m = \dfrac{3}{20}$; for this data set, an increase in temperature of 20° correlates to an increase in snowfall of 3 inches.

3. $y = \dfrac{3}{20}x - \dfrac{3}{2}$

Station 4

Students will be given graph paper, a ruler, and a measuring stick. Students will measure their heights using the measuring stick. They will record this data in a table along with their birth date. They will create a scatter plot of this data. Then they will find the equation of the line of best fit. They will use this equation to predict points on the line that are inside and outside of the data set.

Answers

1–7. Answers will vary.

8. No, you can't predict the height of a person based on their birth date. Your birth date has nothing to do with how tall you are going to be.

Materials List/Setup

Station 1	ruler; graph paper
Station 2	ruler; graph paper
Station 3	ruler; graph paper
Station 4	ruler; graph paper; measuring stick

Algebra I Station Activities for Common Core State Standards

Discussion Guide

To support students in reflecting on the activities and to gather some formative information about student learning, use the following prompts to facilitate a class discussion to "debrief" the station activities.

Prompts/Questions

1. What is a scatter plot?

2. What is the line of best fit on a scatter plot?

3. How do you find the slope of the line of best fit?

4. How do you find an equation for the line of best fit?

5. On a scatter plot, how can you predict data points that weren't in the original data set?

Think, Pair, Share

Have students jot down their own responses to questions, then discuss with a partner (who was not in their station group), and then discuss as a whole class.

Suggested Appropriate Responses

1. A scatter plot is a graph of data points for two or more variables showing their relationships.

2. The line of best fit is a straight line that best represents the data on the scatter plot.

3. Select two data points that best represent the data. Use the points to find the slope.

4. Select two data points that best represent the data. Use the points to find the slope. Use this slope and one of the data points to find the equation of the line of best fit.

5. Find the equation for the line of best fit. Substitute x into this equation to find y and vice versa.

Possible Misunderstandings/Mistakes

- Not understanding that the line of best fit depicts the correlation and slope of the set of data points

- Mistakenly applying the correlation found in the data set to all data sets

- Not understanding the difference between positive, negative, and no correlation

- Using data points that don't best represent the data to find the slope

Statistics and Probability
Set 1: Line of Best Fit

Station 1

You will be given a ruler and graph paper. Use them along with the problem scenario and table below to answer the questions.

Theresa wanted to find a relationship between the number of hours she studied before a math test and her test score. She used a table to track the number of hours she studied and the resulting score on her math test.

Hours studied	Math test score (%)
2	75
4	88
1	65
2	85
8	91
9	97
3	86

1. Work together to graph these ordered pairs. Do not connect the points.

    What type of graph have you created?

2. What happens to the test scores as Theresa increases the number of hours she studies?

    This relationship is called a correlation between the two variables.

3. Using your graph, what type of correlation was there between the number of hours Theresa studied and her test score? Explain your reasoning.

    Using your graph, does this data set have a positive or negative slope? Explain your reasoning.

continued

4. Can you apply the correlation you found in this data set to all of Theresa's subjects, such as English and history? Explain your reasoning.

5. On your graph paper, construct a scatter plot that has a positive correlation.

6. On your graph paper, construct a scatter plot that has a negative correlation.

7. On your graph paper, construct a scatter plot that has no correlation.

Statistics and Probability
Set 1: Line of Best Fit

Station 2

At this station, you will find a ruler and graph paper. The scatter plot below represents a set of data.

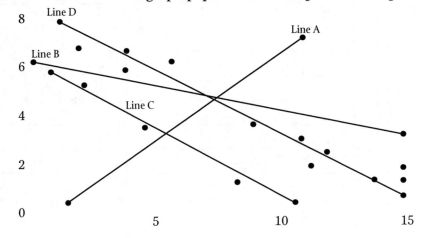

1. Does Line A, B, C, or D represent the line of best fit? Explain your answer.

For each of the following scatter plots, draw the line of best fit and describe its slope.

2.

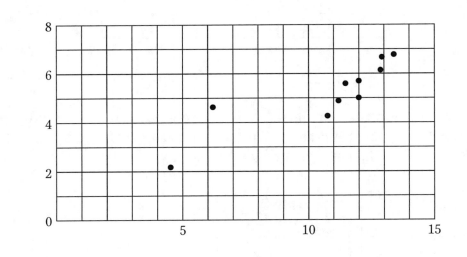

Slope: _____

Statistics and Probability
Set 1: Line of Best Fit

3.

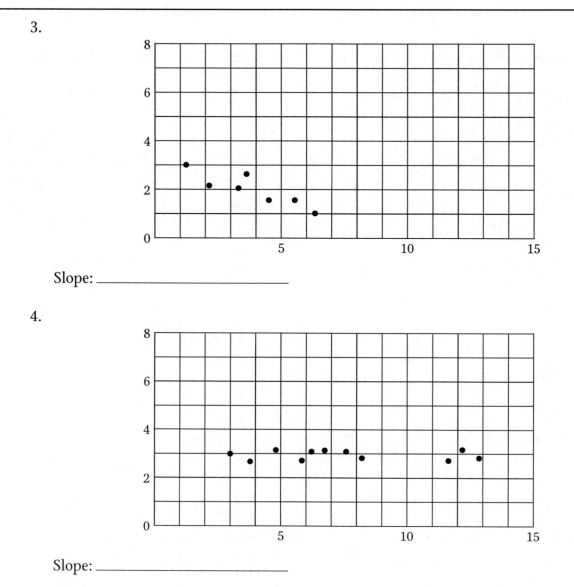

Slope: _____

4.

Slope: _____

5. In the table below, fill in the numerical value for your birthday (month and day) for each member in your group. For example, a person born February 29 would fill in 2 for the month and 29 for the day. Then use your graph paper to create a scatter plot of this data and find the line of best fit.

Month	Day

Statistics and Probability
Set 1: Line of Best Fit

Station 3

At this station, you will find a ruler and graph paper. The data table below represents the average daily temperature versus the daily amount of snowfall for Michigan during a two-week period.

Average daily temperature (°F)	30	31	28	17	34	16	10	19	27	28	24	18	31	30	29
Average daily snowfall (inches)	3	3	4	2	3	0	0	0	1	4	2	2	3	2	5

1. As a group, use your graph paper and ruler to create a scatter plot and line of best fit for this data set.

   Which two data points best represent the data set? Explain your answer.

2. Use these two points to find the slope of the line of best fit. Show your work.

   For this data set, what does the slope represent in terms of temperature and inches of snow?

3. Work together to use these two points and slope to write an equation for the line of best fit. Write your equation in slope-intercept form. Show your work.

© 2011 Walch Education

Statistics and Probability
Set 1: Line of Best Fit

Station 4

At this station, you will find graph paper, a ruler, and a measuring stick. As a group, use the measuring stick to find the height (in inches) of each person in your group.

1. In the table below, fill in the day of the month each person was born and his or her height.

Birthday (day)	Height (inches)

2. On your graph paper, create a scatter plot of the data points in the table. Let *x* represent the birthday values from the table and *y* represent the height values.

3. Which two data points best represent the line of best fit?

4. Find the equation for the line of best fit using these two data points. Show your work in the space below. Write the equation in slope-intercept form.

5. How can you predict a new data point using this equation?

6. If $x = 24$, what value do you predict *y* will be?

7. If $x = 6$, what value do you predict *y* will be?

8. In the real world, can you predict the height of a person based on his or her birth date? Why or why not?

Statistics and Probability

Set 2: Probability

Goal: To provide opportunities for students to develop concepts and skills related to the counting principle and simple and compound probabilities for both independent and dependent events

Common Core Standards

Statistics and Probability: Conditional Probability and the Rules of Probability

Understand independence and conditional probability and use them to interpret data.

S-CP.1. Describe events as subsets of a sample space (the set of outcomes) using characteristics (or categories) of the outcomes, or as unions, intersections, or complements of other events ("or," "and," "not").

S-CP.2. Understand that two events A and B are independent if the probability of A and B occurring together is the product of their probabilities, and use this characterization to determine if they are independent.

S-CP.3. Understand the conditional probability of A given B as $P(A \text{ and } B)/P(B)$, and interpret independence of A and B as saying that the conditional probability of A given B is the same as the probability of A, and the conditional probability of B given A is the same as the probability of B.

Student Activities Overview and Answer Key

Station 1

Students will be given 3 pieces of red yarn, 3 pieces of blue yarn, and tape. Students will be given five index cards that have the following written on them: "chicken," "tuna," "white," "wheat," and "Italian." Students will use the index cards and yarn to model the counting principle of independent events.

Answers

1. No, because there are three types of bread available.

2. chicken, white bread; chicken, wheat bread; chicken, Italian bread

3. tuna, white bread; tuna, wheat bread; tuna, Italian bread

4. multiplication; (2 types of meat)(3 types of bread) = 6 possibilities

5. (4)(3)(8)(4) = 384 different meals

6. no; independent

Station 2

Students will be given index cards with the following subjects written on them: "math," "science," "English," "history," "physical education," and "computer lab." Students will arrange the index cards to create a class schedule. They will derive the counting principle for dependent events by analyzing how to create all possible class schedules.

Answers

1. math, science, English, history, physical education, computer lab; 6

2. Answers will vary.

3. 5; 5

4. Answers will vary.

5. 4; 4

6. Answers will vary.

7. 3; 3

8. Answers will vary.

9. 2; 2

10. Answers will vary.

11. 1; 1

12. Answers will vary.

13. Multiply $(6)(5)(4)(3)(2)(1) = 720$ possible schedules

14. Dependent, because you can only use each class once in the schedule.

Station 3

Students will be given a number cube and fair coin. Students will use the number cube to model simple probability. Then they will use the fair coin to model simple probability. They will explore mutually exclusive events and provide real-world examples.

Answers

1. Answers will vary.

2. equal

3. 6

4. 1, 2, 3, 4, 5, 6

5. $P(5) = 1/6$; $P(6) = 1/6$

6. $P(>4) = P(5) + P(6) = 1/6 + 1/6 = 1/3$; P(even number) $= 1/2$

7. H, T; 2

8. 1/2

9. $1/2 + 1/2 = 1/4$; they are independent events.

10. casino and carnival games

Station 4

Students will be given a bag of 4 marbles that are red, green, yellow, and blue. They will also be given a fair coin. Students will use the marbles to model independent and dependent events. Then they will use the marbles and fair coin to model compound probability and mutually exclusive events.

Answers

1. 4

2. Answers will vary.

3. P(Color drawn) $= 1/4$

4. red and green

5. red and yellow

6. red and blue

7. green and yellow

8. green and blue

9. yellow and blue

10. Answers will vary.

11. P(Pair of marble colors) $= 1/6$

12. P(Green) $= 1/4$

13. H, T; 1/2

14. $P(T) = 1/2$

15. independent

16. Green/T; Green/H; Red/T; Red/H; Blue/T; Blue/H; Yellow/T; Yellow/H

17. 8; $P(G/T) = 1/8$

18. $P(G/T) = P(G) \bullet P(T) = 1/4 \bullet 1/2 = 1/8$

Materials List/Setup

Station 1 3 pieces of red yarn; 3 pieces of blue yarn; tape; five index cards that have the following written on them:

"chicken," "tuna," "white," "wheat," "Italian"

Station 2 six index cards with the following written on them:

"math," "science," "English," "history," "physical education," "computer lab"

Station 3 number cube; fair coin

Station 4 bag of 4 marbles that are red, green, yellow, and blue; fair coin

Discussion Guide

To support students in reflecting on the activities and to gather some formative information about student learning, use the following prompts to facilitate a class discussion to "debrief" the station activities.

Prompts/Questions

1. How do you use the fundamental counting principle to find the number of possibilities of independent events?

2. How do you use the fundamental counting principle to find the number of possibilities of dependent events?

3. How do you find the simple probability of independent events?

4. How do you find the compound probability of independent events?

5. How do casinos and carnivals rely on probability to make money?

Think, Pair, Share

Have students jot down their own responses to questions, then discuss with a partner (who was not in their station group), and then discuss as a whole class.

Suggested Appropriate Responses

1. Multiply together the number of possibilities for each event.

2. Multiply (number of possibilities) times (number of possibilities – 1) and so on.

3. The simple probability of one event equals the number of true outcomes/number of equally likely outcomes. The simple probability of two independent events equals the probability of the first event plus the probability of the second event.

4. The compound probability of two independent events is the probability of the first event multiplied by the probability of the second event.

5. Casinos and carnivals set up games that have the odds in their favor.

Possible Misunderstandings/Mistakes

- Not understanding the difference between independent and dependent events

- Forgetting to subtract 1 from the number of possibilities each time you perform multiplication using the counting principle for dependent events

- Not realizing that the simple probability of two or more independent events is found by adding the individual probabilities

- Not realizing that the compound probability of two or more independent events is found by multiplying each individual probability by the others

Algebra I Station Activities for Common Core State Standards

Statistics and Probability
Set 2: Probability

Station 1

You will be given 3 pieces of red yarn, 3 pieces of blue yarn, and tape. You will also be given five index cards that have the following written on them: "chicken," "tuna," "white," "wheat," and "Italian." Use these materials and the problem scenario below to answer the questions.

> Jon is at the sandwich shop ready to eat lunch, but can't decide what sandwich to order. He wants to get either chicken or tuna. He has narrowed down the type of bread to white, wheat, or Italian.

1. Is Jon choosing between two sandwiches? Why or why not?

2. As a group, determine how many different chicken sandwiches he can pick from by taping the 3 pieces of red yarn to the "chicken" index card and each end of the yarn to the "white," "wheat," or "Italian" cards.

   What sandwich options have you created?

3. Repeat this process with the "tuna" index card and the blue yarn.

   What sandwich options have you created?

4. Based on the number of different sandwiches you have created, what operation (+, –, •, or ÷) can you use to find the number of sandwiches Jon can choose from? Explain your answer.

continued

Statistics and Probability
Set 2: Probability

Use your observations from problems 1–4 to help you answer the next question.

5. A local restaurant offers a four-course meal that includes soup, salad, an entrée, and a dessert for $19.95. You can choose from 4 soups, 3 salads, 8 entrées, and 4 desserts.

    How many different four-course meals can you create? Show your work.

6. Does your choice of entrée depend on your choice of salad? (yes or no) _____

    This means the events are _____ events.

Statistics and Probability
Set 2: Probability

Station 2

You will be given six index cards with the following written on them:

"math," "science," "English," "history," "physical education," "computer lab"

Use these index cards and the problem scenario below to answer the questions that follow.

The computer at school is creating a class schedule for a student named Elena. There are 6 class periods in the day.

1. List the possible classes Elena could have first period. How many possible classes did you list?

    _____

2. Select one of the index cards as Elena's first-period class. Place the index card underneath this paper.

    What class did you choose? _____

3. Look at your index cards. How many index cards are left? _____

    This means there are how many choices for Elena's second-period class?

    _____

4. Select one of the remaining index cards as Elena's second-period class. Place the index card underneath this paper.

    What class did you choose? _____

5. Look at your index cards. How many index cards are left? _____

    This means there are how many choices for Elena's third-period class?

    _____

continued

6. Select one of the remaining index cards as Elena's third-period class. Place the index card underneath this paper.

   What class did you choose? _____

7. Look at your index cards. How many index cards are left? _____

   This means there are how many choices for Elena's fourth-period class?

   _____

8. Select one of the remaining index cards as Elena's fourth-period class. Place the index card underneath this paper.

   What class did you choose? _____

9. Look at your index cards. How many index cards are left? _____

   This means there are how many choices for Elena's fifth-period class?

   _____

10. Select one of the remaining index cards as Elena's fifth-period class. Place the index card underneath this paper.

    What class did you choose? _____

11. Look at your index cards. How many index cards are left? _____

    This means there are how many choices for Elena's sixth-period class?

    _____

12. Select the last remaining index card as Elena's sixth-period class. Place the index card underneath this paper.

    What class did you choose? _____

continued

Statistics and Probability
Set 2: Probability

13. You created one possible class schedule for Elena. How can you use your answers in problems 1, 3, 5, 7, 9, and 11 to find ALL the possible class schedules Elena could have?

14. Were the choices you selected for each class period dependent on your choices for the other class periods? Why or why not?

Station 3

You will be given a number cube and a fair coin. You will use these tools to explore simple probability.

1. As a group, roll the number cube. What number did you roll? _____

2. Did the number you rolled have less, equal, or more of a chance of being rolled than the other numbers on the number cube? Explain your answer.

3. How many numbers are on the number cube? _____

4. List these numbers: _____

   In probability, these numbers are known as your sample space or possible outcomes.

5. What is the probability of rolling a 5 on your number cube? Explain your answer.

   What is the probability of rolling a 6 on your number cube? Explain your answer.

6. How can you use your answers in problem 5 to find the probability P of rolling a number > 4 on your number cube?

continued

Statistics and Probability
Set 2: Probability

What is the $P(\text{number} > 4)$? _____

What is the $P(\text{even number})$? _____

As a group, examine the coin.

7. What is the sample space of the coin? _____

    How many possible outcomes can you have when you flip the coin? _____

8. What is the probability of tossing the coin heads-up? _____

9. What is the probability of tossing the coin heads-up, then tossing it heads-up again? Explain your answer.

10. What are some examples of simple probability used in the real world?

Statistics and Probability
Set 2: Probability

Station 4

You will be given a bag of 4 marbles that are red, green, yellow, and blue. You will also be given a fair coin. You will use these tools to explore compound probability. Work as a group to answer the questions.

1. Place the marbles in the bag. How many different colored marbles are in the bag?

    _____

2. Pick one marble from the bag without looking. What color marble did you choose?

    _____

3. How can you use your answers from problems 1 and 2 to find the probability of choosing a marble of the color you found in problem 2?

Place the marble back in the bag. What are the possible different pairs of marble colors in the bag? A possible pair has been given for problem 4 as a model.

4. _____red and green_____
5. _____
6. _____
7. _____
8. _____
9. _____

10. Pick two marbles from the bag without looking. What pair of marble colors did you choose?

    _____

continued

Statistics and Probability
Set 2: Probability

11. How can you use your answers from problems 4–9 to find the probability of choosing the pair of marble colors you found in problem 10? (*Hint*: These are dependent events.)

Place the marbles back in the bag.

12. What is the probability of choosing a green marble? _____

13. Examine your fair coin. List the possible outcomes that you could get if you tossed the coin.

    _____

    The probability of each outcome is _____ .

14. What is the probability of tossing the coin tails-up? _____

15. Are these two events independent or dependent on each other? Explain your answer.

16. List the sample space for finding the probability of choosing a green marble AND then tossing a coin tails-up. *P*(green/tail).

17. How many outcomes did you find in problem 16? _____

    What is *P*(green/tail)? _____

18. What is a faster way to find the *P*(green/tail) than listing the entire sample space?

Statistics and Probability

Set 3: Data Displays

Goal: To provide opportunities for students to develop concepts and skills related to mean, median, and mode for stem-and-leaf plots and box-and-whisker plots, including the interquartile range

Common Core Standards

Statistics and Probability: Interpreting Categorical and Quantitative Data

Summarize, represent, and interpret data on a single count or measurement variable.

S-ID.1. Represent data with plots on the real number line (dot plots, histograms, and box plots).

S-ID.2. Use statistics appropriate to the shape of the data distribution to compare center (median, mean) and spread (interquartile range, standard deviation) of two or more different data sets.

S-ID.3. Interpret differences in shape, center, and spread in the context of the data sets, accounting for possible effects of extreme data points (outliers).

Student Activities Overview and Answer Key

Station 1

Students will be given two number cubes. Students will use the number cubes to create 11 two-digit numbers. Then they will create a stem-and-leaf plot to represent this data. They will find the mean, median, and mode of the data.

Answers

1. Answers will vary. Possible answers include:

Unordered	Ordered
16	11
23	14
14	16
45	22
34	23
23	23
11	34
22	45
45	45
45	45
51	51

2. first digit

3. second digit

4. stem

5.

Stem	Leaf
1	1 4 6
2	2 3 3
3	4
4	5 5 5
5	1
6	

6. The number cube only has digits 1–6.

7. Mode; answers will vary. Possible answer: 45

8. Median; answers will vary. Possible answer: 23

9. Mean; answers will vary. Possible answer: 29.91

10. Answers will vary. Possible answer: student test scores

Station 2

Students will be given four number cubes. Students will divide into two groups, each with two number cubes, which they will use to create two-digit numbers. Then they will work as one group to construct a back-to-back stem-and-leaf plot. They will find the mean, median, and mode of this back-to-back stem-and-leaf plot. They will give a real-world example of a back-to-back stem-and-leaf plot.

Answers

1. Answers will vary. Possible answer:

Group 1	Group 2
44	22
21	21
15	16
16	11
24	36
61	41
55	42

2. stem

3. Answers will vary. Possible answer:

Leaf	Stem	Leaf
5 6	1	1 6
1 4	2	1 2
	3	6
4	4	1 2
5	5	
1	6	

4. 14 numbers, because of both data sets

5. Answers will vary. Possible answer: 21 and 16; mode

6. Answers will vary. Possible answer: 23; median

7. Answers will vary. Possible answer: 30.4; mean

8. sports scores of two teams

Station 3

Students will be given nine index cards with the following numbers written on them: 52, 49, 69, 44, 88, 80, 68, 49, 90. Students will also be given graph paper. Students work together to find the minimum, maximum, mode, median, and sub-medians of the data set. They will find the interquartile range and look for outliers. Then on graph paper, they will construct a box-and-whisker plot to represent the data set. They explain when a box-and-whisker plot would be useful to analyze data.

Answers

1. 44

2. 90

3. 49; mode

4. 68; 68 is the median because there are four numbers below and above it in the data set; 44, 49, 49, 52; $\frac{(49 + 49)}{2} = 49$; 69, 80, 88, 90; $\frac{(80 + 88)}{2} = 84$

5. 49; it is the median of the lower numbers.

6. Median: 68

7. 84; it is the median of the upper numbers.

8. 44, because it is the minimum data point; 35; no, because $44 > 49 - 1.5(35)$

9. 90, because it is the maximum data point; no, because $90 < 84 + 1.5(35)$

10.

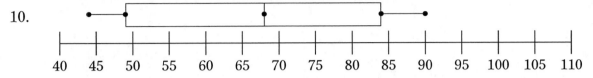

11. Answers will vary. Possible answer: Box-and-whisker plots show how the data is dispersed over the data set.

Station 4

Students will be given graph paper and a ruler. Students will create two sets of data. They will find the quartiles, interquartile range, and medians for each data set. Then they will create parallel box-and-whisker plots of the data sets. They will explain when parallel box-and-whisker plots are useful.

Answers

1–5. Answers will vary.

6. Make sure the box-and-whisker plot is graphed correctly.

7–11. Answers will vary.

12. Make sure the box-and-whisker plot is graphed correctly and is on the same graph as problem 6.

13. Parallel box-and-whisker plots make it easy to compare two sets of data.

Materials List/Setup

Station 1   two number cubes

Station 2   four number cubes

Station 3   graph paper; nine index cards with the following numbers written on them:

52, 49, 69, 44, 88, 80, 68, 49, 90

Station 4   graph paper; ruler

Discussion Guide

To support students in reflecting on the activities and to gather some formative information about student learning, use the following prompts to facilitate a class discussion to "debrief" the station activities.

Prompts/Questions

1. What is the mode of a data set?

2. How do you find the median of a data set?

3. How do you find the mean of a data set?

4. How do you find the interquartile range in a box-and-whisker plot?

5. How do you determine if there are outliers in a box-and-whisker plot?

6. When are stem-and-leaf plots useful?

7. When are parallel box-and-whisker plots useful?

Think, Pair, Share

Have students jot down their own responses to questions, then discuss with a partner (who was not in their station group), and then discuss as a whole class.

Suggested Appropriate Responses

1. The mode is the number that occurs most often in the data set.

2. The median is the number that has the same amount of numbers above and below it in the data set.

3. The mean is the sum of the data set divided by the number of numbers in the data set.

4. Interquartile range $= Q_3 - Q_1$

5. An outlier is a number less than $Q_1 - 1.5(\text{interquartile range})$ or a number greater than $Q_3 + 1.5(\text{interquartile range})$.

6. Stem-and-leaf plots are an easy way to analyze specific points in one or more data sets.

7. Parallel box-and-whisker plots let you compare two data sets and compare their dispersion.

Possible Misunderstandings/Mistakes

- Not writing data sets in numerical order before creating stem-and-leaf plots and box-and-whisker plots

- Confusing the stem with the leaf in stem-and-leaf plots

- Not using the stem as the central component of back-to-back stem-and-leaf plots

- Not identifying outliers in data sets for box-and-whisker plots

- Incorrectly finding Q_1, Q_2, and/or Q_3 in box-and-whisker plots

Statistics and Probability
Set 3: Data Displays

Station 1

You will be given two number cubes. As a group, roll the first number cube and write the result on the first line below. Then roll the second number cube and write the result on the second line below.

_____ _____

You have created a two-digit number. Repeat this process until you have created 11 two-digit numbers. As you create each two-digit number, write it in the "Unordered" table on the left, below. Don't forget to include the first two-digit number you created.

Then, organize the same 11 numbers in numerical order. Write the numbers in the "Ordered" table on the right.

1.

Unordered

Ordered

You can represent this ordered data in a stem-and-leaf plot.

2. The "stem" is like the stem of a flower. Look at your table of values. Which digit do you think represents the "stem" of each number?

_____

continued

3. The "leaf" is like the leaf of a flower which extends off the stem. Look at your table of values. Which digit do you think represents the "leaf" of each number?

   _____

4. If your data table had numbers 44, 48, 49, and 42, then they would have the same

   _____ .

5. Write each digit from your table in the stem-and-leaf plot below. (The possible stems have been given for you.)

Stem	Leaf
1	
2	
3	
4	
5	
6	

6. Why were the possible stems only the numbers 1–6?

You can find the mean, median, and mode of the data in the stem-and-leaf plot.

7. Examine your stem-and-leaf plot. Which number occurs most often? _____

   Is this number the mean, median, or mode? Explain your answer.

8. Examine your stem-and-leaf plot. Which number is in the middle of all the other numbers when the numbers are written in numerical order?

_____

Is this number the mean, median, or mode? Explain your answer.

9. What number do you get if you add up all the numbers in your stem-and-leaf plot and divide by 11?

_____

Is this the mean, median, or mode? Explain your answer.

10. What is an example of real-world data you could display in a stem-and-leaf plot?

Statistics and Probability
Set 3: Data Displays

Station 2

You will be given four number cubes. Your group will split up into two small groups. Each small group will have two number cubes. Each small group will roll the number cubes to create 7 two-digit numbers.

1. Each group will record their two-digit numbers in the tables below.

Group 1	Group 2

Now your two small groups will combine and work as one group to represent this data in a "back-to-back" stem-and-leaf plot.

2. Will both data sets have the "stem" or "leaf" in common? Explain your answer.

3. Create a "back-to-back" stem-and-leaf plot that uses both sets of data:

Leaf	Stem	Leaf

continued

Statistics and Probability
Set 3: Data Displays

You can find the mean, median, and mode of the data in a back-to-back stem-and-leaf plot.

4. Examine your stem-and-leaf plot. How many numbers would you have to analyze if you were going to find the mean, median, and mode? Explain your answer.

5. Which number occurs most often? _____

   Is this number the mean, median, or mode? Explain your answer.

6. Which number is in the middle of all the other numbers when the numbers are written in numerical order?

   _____

   Is this number the mean, median, or mode? Explain your answer.

7. What number do you get if you add up all the numbers in your stem-and-leaf plot and divide by the total number of numbers?

   _____

   Is this number the mean, median, or mode? Explain your answer.

8. What is an example of two real-world data sets you could display in a back-to-back stem-and-leaf plot?

Statistics and Probability

Set 3: Data Displays

Station 3

You will be given nine index cards with the following numbers written on them:

52, 49, 69, 44, 88, 80, 68, 49, 90

You will also be given graph paper.

As a group, arrange the index cards in numerical order.

1. Which number is the smallest number in the data set? _____

2. Which number is the largest number in the data set? _____

3. Which number occurs most often in this data set? _____

   What is the name for this number? _____

4. Which number is the median in this data set? Explain your answer.

   Which four numbers are less than the median for this data set?

   _____

   What is the median of these four numbers? Show your work.

   _____

continued

Which four numbers are greater than the median for this data set?

_____

What is the median of these four numbers? Show your work.

_____

You can create a box-and-whisker plot to represent your data set. A box-and-whisker plot looks like the following:

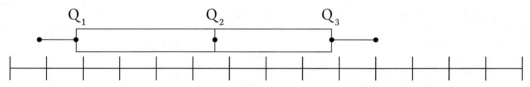

Examine the box-and-whisker plot.

5. Which number from your data set do you think represents the point Q_1? Explain your answer.

6. Which number from your data set do you think represents the point Q_2? Explain your answer.

7. Which number from your data set do you think represents the point Q_3? Explain your answer.

continued

8. Which number from your data set do you think represents the left end point of the line? Explain your answer.

This number may be an outlier. If that is the case, then it can't be the left end point of the line on the box-and-whisker plot.

Use the interquartile range to see if this minimum number is an outlier.

Interquartile range $= Q_3 - Q_1 =$ _____

An outlier in this direction will be any point on a number line that is less than $Q_1 - 1.5$(interquartile range).

Is the minimum number in the data set an outlier? Why or why not?

9. Which number from your data set do you think represents the right end point of the line? Explain your answer.

This number may be an outlier. If that is the case, then it can't be the right end point of the line of the box-and-whisker plot.

An outlier in this direction will be any point on a number line that is greater than $Q_3 + 1.5$(interquartile range).

Is the maximum number in the data set an outlier? Why or why not?

continued

10. On your graph paper, construct the box-and-whisker plot for your data set.

11. How are box-and-whisker plots useful in analyzing data?

Algebra I Station Activities for Common Core State Standards
© 2011 Walch Education

Statistics and Probability
Set 3: Data Displays

Station 4

You will be given graph paper and a ruler. You will use these materials, along with data on the birthdates of your group members and your group members' mothers, to answer the questions and construct parallel box-and-whisker plots.

A box-and-whisker plot looks like this:

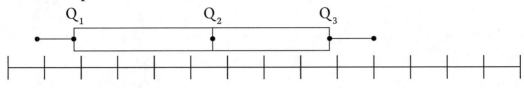

In the table below, write the number of the day of the month each student in your group was born. For example, if a student was born on February 29, write "29" in the table. Write these numbers in numerical order.

Your Birthday (Day)

1. What is the median of this data set? _____

   This will be Q_2 in your box-and-whisker plot.

2. What is the median of the numbers that are *less than* your answer from problem 1?

   _____

   This will be Q_1 in your box-and-whisker plot.

continued

3. What is the median of the numbers that are *greater than* your answer from problem 1?

   _____

   This will be Q_3 in your box-and-whisker plot.

4. What is the interquartile range of this data set? Show your work. (*Hint*: interquartile range = $Q_3 - Q_1$)

Find out if there are any outliers in your data set:

   A number is an outlier if it is less than $Q_1 - 1.5$(interquartile range).

   A number is an outlier if it is greater than $Q_3 + 1.5$(interquartile range).

5. Are there any outliers in this data set? If so, what are they?

   _____

6. On your graph paper, graph a box-and-whisker plot that represents the data you gathered on your group members' birthdays.

   Now, you will follow the same process to create a box-and-whisker plot for the birthdays of your group members' mothers.

   In the table below, give the number of the day of the month on which each group member's mother was born. For example, if a group member's mother was born on May 14, write "14" in the table. Write these numbers in numerical order.

Mother's Birthday (Day)

continued

7. What is the median of this data set? _____

   This will be Q_2 in your box-and-whisker plot.

8. What is the median of the numbers that are *less than* your answer from problem 7?

   _____

   This will be Q_1 in your box-and-whisker plot.

9. What is the median of the numbers that are *greater than* your answer from problem 7?

   _____

   This will be Q_3 in your box-and-whisker plot.

10. What is the interquartile range of this data set? Show your work. (*Hint*: interquartile range = $Q_3 - Q_1$)

Find out if there are any outliers in your data set:

   A number is an outlier if it is less than $Q_1 - 1.5$(interquartile range).

   A number is an outlier if it is greater than $Q_3 + 1.5$(interquartile range).

11. Are there any outliers in this data set? If so, what are they?

   _____

continued

12. On the same sheet of graph paper you used for problem 6, graph a box-and-whisker plot that represents the data you gathered on your mothers' birthdays.

13. Why is it useful to construct parallel box-and-whisker plots?